建筑重现
The Palimpsest Effect

[丹]BIG建筑事务所等 | 编
司炳月 李一同 孙彤彤 | 译

大连理工大学出版社

004 他们叫作"变量人"_ Adrian Friend

新项目

008 Bugye植物园：思惟园 _ Seung H-Sang

독与群：思惟园与承孝相的建筑 _ Hyungmin Pai

感受实用主义

070 感受实用主义：重新思考公共设施建筑 _ Nelson Mota

078 哥本丘 _ BIG

094 坪山阳台 _ NODE Architecture & Urbanism

110 索勒德加德净水厂 _ Henning Larsen

124 穆滕茨净水厂 _ Oppenheim Architecture

建筑重现

142 建筑重现 _ Richard Ingersoll

148 Z33当代艺术、设计与建筑美术馆 _ Francesa Torzo Architetto

Interview _ Francesa Torzo + Herbert Wright

170 雅各比工作室 _ David Chipperfield Architects

188 洛桑州立美术馆 _ Barozzi Veiga

204 科隆学校附属建筑和住房 _ LRO Lederer Ragnarsdóttir Oei Architekten

220 建筑师索引

004 Calling the "Variable Man"_Adrian Friend

New Projects

008 Bugye Arboretum "Sayuwon"_Seung H-Sang

Alone and Together: Sayuwon and the Architecture of Seung H-Sang_Hyungmin Pai

Perceptive Utilitarianism

070 Perceptive Utilitarianism: Rethinking the Architecture of Public Amenities_Nelson Mota

078 CopenHill_BIG

094 Pingshan Terrace_NODE Architecture & Urbanism

110 Solrødgård Water Treatment Plant_Henning Larsen

124 Muttenz Water Purification Plant_Oppenheim Architecture

The Palimpsest Effect

142 The Palimpsest Effect_Richard Ingersoll

148 Z33 House for Contemporary Art, Design and Architecture_Francesa Torzo Architetto

Interview_Francesa Torzo + Herbert Wright

170 Jacoby Studios_David Chipperfield Architects

188 Cantonal Museum of Fine Arts_Barozzi Veiga

204 School Annex and Housing, Cologne-Lindenthal_LRO Lederer Ragnarsdóttir Oei Architekten

220 Index

他们叫作"变量人"
Calling the "Variable Man"

Adrian Friend

"这家伙与众不同。没有他修不好的东西;他简直无所不能。各类事实积累成为学问,构成科学,但他做起事来既不靠学问,也不靠科学。他对这两个概念一无所知。脑子里根本没有'学习'这两个字。他全靠直觉行动,他的魔力全在他的手中。他是个多面手,快看这双手!这双手像是画家,又像是艺术家。他的双手还像是利刃,从我们的生活中划过。"[1]

科幻小说作家菲利普·K.迪克短篇集《变量人》中的主人公意外地穿越到了这样一个时代,各个知识体系是独立的,人们专精于某一门知识。而这位"变量人"则用自己的双手游走在各个知识版图中,他并不专精于某一学科,但其中的流程和技巧他却了然于胸。他从自己的修理铺穿越到一所实验室,用这里的工具去解决他所擅长的问题。

我相信科幻小说中的"变量人"终于迎来了春天,并且能够胜任21世纪的工作,这个"变量人"就是建筑师。近一百年前,在建筑师这个职业刚刚兴起之时,为了能够控制建筑师的心理,在行业中曾发生过不少争议,从那时起,就流传着这样一句话,"建筑师是专才之海里的通才"。1825年,通过律师协会,律师成为正式的职业;1858年,通过英国医学总会(GMC),医生成为正式的职业;1932年,通过建筑师注册委员会(ARB),建筑师也成了正式的职业。

在最开始,"建筑"是匠人们手中的窗饰和图纸,到了20世纪中叶,"建筑"成了设计新的城市,而在

"This man is different. He can fix anything; he can do anything. He doesn't work with knowledge, with science - the classified accumulation of facts. He knows nothing. It's not in his head, a form of learning. He works by intuition - his power is in his hands, not his head. Jack-of-all-trades. His hands! Like a painter, an artist. And with his hands, he cuts across our lives like a knife-blade."[1]

SciFi writer Philip K. Dick's "Variable Man" is accidentally brought forwards to a time when knowledge is compartmentalized, people highly specialized. He crosses these boundaries by working with his hands, knowing about processes and techniques rather than specific disciplines. Parted from his Fixit cart, he lands in a lab and sets about using their tools to solve a problem adjacent to the one suggested to him.

I believe the "variable man" of sci-fi has finally arrived and can be assigned to a 21st century profession, that is the architect – a generalist in a sea of specialists as was a phrase that was always part of the internal struggle for control of the architect psyche going as far back the dawn of the profession nearly one hundred years ago. The register of professions through the Architects Registration Board (ARB) in 1932 followed medicine and the General Medical Council (GMC) in 1858 and Law and the Law Society in 1825.

Collaboration is and has always been at the heart of architecture, originally born out of crafts people's tracery and drawings and then later when emerged in mid-twentieth century planning new cities – collaboration with other rising professions has always been at the heart of the architect – of all the professions they have always been the most "variable" and encompassing. The rise of inter and multidisciplinary architecture courses was cemented by the important conference on architectural education that was held at Oxford in 1958 where, "it was decided as a matter of future policy that schools of architecture should be within universities whenever possible." At that time there were "thirty-one British Schools, and thirteen are departments of universities, the

此期间，合作一直是"建筑"的核心概念，现在仍是如此。同其他不断兴起的职业合作也一直是"建筑师"工作的核心——但在所有职业中，建筑师是最"多面化"的，建筑师是最具"包容性"的。1958年，牛津大学（Oxford）举办了一场重要的建筑学教育大会，大会决定"大学应在未来尽可能地开设建筑学院"，这场会议对跨领域建筑学课程和多学科建筑学课程的发展起到了巩固作用。当时，"英国有31所建筑学院，其中有13所由大学开设，在其余的18所里，除了独立的建筑联盟学院，剩余的建筑学院都开设在艺术学院或理工学院。尽管能开设的课程十分有限，但这为组织多学科建筑学课程提供了很好的机会"。²

英国大多数的建筑学教育研究都将1958年的这场大会视为英国建筑学教育的转折点。"所谓的多学科教育是当今大环境需求的产物，需要各个行业的密切合作。英国伦敦大学学院最近决定将几个工程学和建筑科学分支合并为一个部门，起名为'环境研究学院'，这是我们当今时代的典型标志，也为教职员工交流和项目合作提供了显而易见的机会。"³在大会上，建筑师莱斯利·马丁表示："若是建筑学教育打算在专业实践中融入现实主义，那最好的开展方式有两种，可以任选其一：第一种是在课程体系内开设实践项目，第二种是全日制学习和实践学习交替进行，又称'三明治模式'，这种模式更加流行。因此英国皇家建筑师协会（RIBA）对建筑师的认证分为了1、2、3共三个部分。"⁴

最初，大学里建筑学教育的目的是让建筑师拥有多样化的技能，同时，大学还成立了建筑环境学院，帮助建筑师和其他专业更好地进行沟通，但在2008年，牛津会议召开50周年之际，许多受访学院的高层人士都忘记了这一最初的目标。布雷特·斯蒂尔是建筑联盟学院的院长，他指出："自20世纪中叶以来，教育走向了彻底的标准化和专业化。"⁵但他并不是对这一逐渐崛起的概念（学科融合）提出挑战，而是将专业化视为新型知识储备的标准。其他学院已经接受了这一认证标准，而十年后，建筑联盟学院才认识到这一点，也开始接受这一认证标准。

自20世纪中叶以来，专业化开始在全社会兴起。我个人比较看重的是应用艺术和工艺领域，尤其是

remainder, apart from the independent AA school, being in art colleges or polytechnics, which also offer an opportunity of organizing multi-disciplinary courses, though in a more limited range of subjects."²

Most studies of architectural education in the UK regard the 1958 Oxford Conference as a turning point in British architectural education. "What is called multi-disciplinary education is a direct result of the need in present-day circumstances, […] for the various professions to work closely together, and a typical sign of the times was a recent decision by University College London, to unite several branches of engineering and building sciences – into one super department called the School of Environmental Studies, thus providing obvious opportunities for the exchange of staff and programs."³ At the Conference, architect Leslie Martin argued that if "the realism of professional practice was to be provided in an architectural education. It could best be provided in one of two ways: either through in-curriculum live projects, or through a sandwich alternating full-time study with periods in practice. The sandwich model would prevail, giving rise to the model of the RIBA Part 1, 2 and 3"⁴.

What many heads of schools interviewed in 2008 on the fiftieth anniversary of the Oxford conference forgot was the origin of architecture being taught within universities was to diversify the skillset of the architect and to connect with other professions also being taught within Schools of the Built Environment that were created. Not to challenge the rising perception as Brett Steele, Director of the AA, said that "education in this country since the mid-20th century has been so thoroughly standardized and professionalised"⁵, but to see the professionalization as a natural canon for new knowledge bases, which the AA recognized ten years later, by succumbing to the same validation systems that the other schools were already being validated under. Professionalism has been on the rise since the mid-20th century in all parts of society. Just

能通过新型数字技术复兴和扩大规模,并成为建筑学一部分内容的美术工艺,也就是说,只有出现衡量能力和质量的新标准,机械工艺才可能出现,这种标准是专业和业余团体共同追求的目标。在第二次世界大战之后,家庭装修和家庭工艺活动在美国和英国风靡一时。因此,出版商开始推出新的刊物,制造商开始设计新的系列产品,这反过来又刺激了人们对装修的需求。在这一时期,购房人数激增,房主的态度也发生了改变。"人们不再为了装修而装修,装修是为了带来更美的生活方式,它成了确立地位和身份的一种方式,也成了自我实现的一种手段。"[6]换句话说,人们渴望变得"专业化",这是一种对技术的探索,也打开了这个行业的大门,使其更加多元化。这反过来开始影响并改变专业标准——我们已经见证了专业人士和业余人士合作运动的兴起,他们使用更加高级的工具,也取得了比专业人士更为理想的效果。[7]

查尔斯·李德彼特(2004年)指出,这种专业人士和业余人士结合的模式让我们的生活结构发生了新的变化,这样的结构是网络化的,需要人们进行合作,这种模式结构简单,并且在很大程度上可以实现自我调节。在科学、药学、战争、政治、教育和福利等领域,专业人士们用自己的知识、权威和制度塑造了20世纪,且在21世纪仍然至关重要。但专业人士和业余人士结合的模式将会成为新的驱动力,创造出新的知识体系、新的结构、新的权威确立依据。当今的建筑师需要了解消防工程和碳中和等新兴的职业,与本职工作互补,他们还要与由专业人士和业余人士组成的队伍建立新的关系。"专业的"业余人士跨越了传统的建筑行业边界,在更加多样化的、可持续的、新的混合实践团体中推动和测试新技术。通常来说,他们的发

looking at an area I hold dear, applied art & craft, especially which can be revived and scaled up to architectural components through new digital tools – i.e. robotic craft has only been possible through emergence of new standards of competency and quality that are shared and aspired to both in the amateur and professional communities. In the period after the Second World War a huge interest in home improvement and home craft activity emerged in Britain and the United States. As a response to this, publishers and manufacturers developed new periodicals and product ranges, which in turn further stimulated demand. In this period home ownership grew enormously, and there was a change in attitude of the householder. "Home improvement was no longer tackled out of necessity but as part of a project leading to more aestheticized lifestyles, a way of establishing status and identity, and a means of self-actualization."[6] In other words the aspiration to be professional was something that was aspired to as a technical quest that arguably opened up the profession to greater diversity. This in turn starts to impact and change professional standards – already witnessed by the rise of the Pro-Am movements which are able to achieve results better than the professionals with more advanced tools.[7]

According to Charles Leadbeater (2004) the Pro-Am's will bring new forms of organization into life, which are collaborative, networked, light on structure and largely self-regulating. Professionals – in science and medicine, war and politics, education and welfare – shaped the twentieth century through their knowledge, authority and institutions. They will still be vital in the 21st century. But the new driving force, creating new streams of knowledge, new kinds of organizations, new sources of authority, will be the Pro-Am's. Today architects need to understand emerging complementary professions such as fire engineering and carbon neutrality. In addition, they also need to establish new relationship with the Pro-Am's. The professional amateurs are fuelling and testing the technology in new sustainable and more diverse blended practice collectives outside the traditional boundaries of the architectural profession. Often, the discoveries and advancements of the Pro-Am's have the power to influence the professional practice of architects. The appropriation of the resources in the most efficient way has always been at the core heart of the profession of architecture, and therefore in the latest life-support systems debates we are only just beginning to

现和取得的进步会影响到建筑师的专业实践。用最有效率的方式分配资源一直都是建筑业的核心活动，因此，在最近的关于生命维持系统的争论中，我们才刚刚开始重拾100年前专业建筑师的技能。

英国皇家建筑师协会可持续未来小组的领导人盖里·克拉克说道："去年，在我最初开始思考开设一门可持续性的课程时，我的目标仅仅是在一系列的分类体系中总结出可持续发展的绝对关键概念。如果你读了这本可持续发展成果指南，就会了解到课程的核心结构。但是随着更多的人付出努力，我最初的想法发生了变化。直到今天，这本指南成了涵盖内容更加广泛的可持续性设计手册。问题在于现在的分类体系和优先目标尚不明确。我认为开始的时候不用做得太多——建筑师们应该从提出问题开始——客户需要的仅仅是建筑，还是需要进一步了解建筑的使用方式和建造过程？如果是前者，我们首先要开始进行改造，在改造结果无法满足客户需求时，再建造新的建筑。我们还需要设计可拆卸和可再利用的建筑。但这并不是我们设计的唯一优先目标，因为我们必须设计出使用绝对最小能源的建筑，它寿命长、能耗低，而且布局宽松！"

这些原则应当成为建筑学教育的核心，并且应当与学生的学习水平保持一致。因为建筑专业的核心一直是以最有效的方式分配资源，因此，我们才开始重拾1958年牛津会议的多元化精神，那就是拥有专业建筑师的技能，并找到连接其他专业的途径。只有"变量人"才能在不同的大环境下寻找正确的发展方向，驾驭不断变化和发展的新兴专业技术，并且与21世纪社会相关的新型混合实践模式相融合。

recapture the professional skills of architects a hundred years ago.
"When I originally was thinking about creating a sustainable curriculum last year, my goal was to just summarise the absolute key concepts of sustainability in a cascading hierarchy," says Gary Clarke, Head of the Royal Institute of British Architects Sustainable Futures Group, "If you read the sustainable outcomes guide this suggests a core structure of a curriculum. However, when more people contributed then the original intention has changed to now a much broader sustainable design handbook. The problem is the hierarchy and priorities are not so clear. I would suggest beginning with doing less – architects should start with the question – does the client need a building or do they need to rationalise their utilisation and processes? If it is found that they need a building, we should start with retrofit first and only if this has been found not to meet their needs do we build new. We should also design for disassembling and re-use. However, that is not the only priority in that we have to design buildings that use the absolute minimum of energy – long life, loose fit and low energy!"
These principles should be at the core of architectural education and should be aligned with student grades as the appropriation of resources in the most efficient way that has always been at the core heart of the profession of architecture, and therefore we are only just recapturing the diverse spirit of the 1958 Oxford Conference with professional architect skills and pathways into other specialisms that only the "variable man" can contextualize and navigate, by variably directed by a growing and emerging technical specificity as part of new blended models of practice that are relevant for 21st century society.

1. Philip K. Dick, *The Variable Man*, revised ed., Marina del Rey, CA, Aegypan, 2011.
2. J. M. Richards, *Architecture*, Cincinnati, OH, David & Charles, 1974.
3. Ibid.
4. L. Martin, *Conference on Architectural Education*, RIBA Journal (June), 1858, pp.279-283.
5. J. Till, S. Brett, I. Borden, M. Echenique, *Questioning the Oxford Agenda*, Building Design, June 13, 2008.
6. A. Jackson, *Men who make: the "flow" of the amateur designer/maker*, Duke University Press, 2011.
7. C. Leadbeater & P. Miller, *The Pro-Am Revolution: How Enthusiasts are changing our economy and society*, Demos, 2004.

山地植物园的沉思
Contemplations in a Mountainous Arboretum

Bugye植物园: 思惟园
Bugye Arboretum "Sayuwon"

Seung H-Sang

人之居也如诗
Poetically Man Dwells

故事开始前，我想说一说韩国大邱市的刘宰成 (Yoo Jae-sung) 主席。他是韩国Taechang钢铁公司的所有者，也是韩国大邱-京畿道区艺术文化发展的赞助商，他很有影响力，却十分低调。

几十年前，他曾经花了一大笔钱买了一批上百年的榠楂树，以防有人将这些树走私到国外。在这件事之后，有人建议他收集全韩国各地稀有的榠楂树，因此他决定创办一个植物园。下定决心之后，他来到庆尚北道军威郡的Bugye买下了一片土地。

十年前，我为他40年的房子设计了扩建的部分，并将其取名为"某轩 (Moheon)"，也就这样和他有了些交情。"Moheon"的意思是"某人的房子"，对于喜欢保持低调的刘宰成来说，这是个很适合他的名字。也许正是出于这个原因，他和我说起了自己的植物园，我们从最开始的概念阶段谈起。我给植物园起名为"思惟园"，也许这是个挺直白的名字，其含义为"反思冥想之地"，也就是"地处Bugye的反思冥想植物园"。孔子曾经说过，"为政必先正名"。"思惟园"这个名字让我们对植物园的特点有了一个清晰的认识。除了植物园以外，里面的建筑设施也必须提供可以进行深度反思和冥想的地方。

最重要的是，我与这片土地的第一次相遇就擦出了火花，因此，我相信这片土地一定可以成为一个反思冥想之处。这里有两个峡谷和三个山脊，位于南面的八公山和北面的道峰山之间。当我第一次登上顶峰时，自然天成的景色映入眼帘，没有任何人工雕饰的痕迹。啊！这里位于大自然的深处，是真正与世隔绝的地方，让我感觉天地之间仿佛只有我自己存在着。如今，这里占地共1 000 000m²，但在视觉效果上看似和邻近的山峰相连，仿佛绵延数亿平方米，给人带来了更为震撼的视觉冲击。

郑永善 (Jung Young-sun) 是一位杰出的韩国景观设计师，我非常信任他，他规划出总体框架，将这片土地划分得非常雅致。日本景观设计师Kawagishi Matsunobu巧妙地设计了刘宰成在大邱市住宅附近的景观，还不时地给我们提出宝贵的意见。在他们的帮助下，植物园独特的景观设计开始成型。

暖阳照进森林深处，野花野草生机盎然，涓涓溪流让鸟儿们流连忘返。移植过来的榠楂树让周围的环境更显柔和静谧，与美丽繁茂的紫薇树形成了鲜明的对比。这里每棵松树的价值都是天文数字，在这片森林中尽情散发着自己的魅力，或许，这些松树正是这里的"魅力之王"。以上种种结合起来，这一全新的景观给人以波澜壮阔之感。

景观和树木是这片土地的主角，因此建筑外观不能太过奢华。从建筑学的角度来说，这座建筑非常低调，其选址经过深思熟虑，最终才成了一个不错的观景之处。因此，我在这里设计的第一座建筑，名为玄庵 (Hyeonam，黑色小屋)，和这里的其他建筑一样，要么低到贴近地面，要么深入到地下。我唯一的愿望是创造出一座仿佛一直存在于此处的建筑。

玄庵是天地之间的一片空间；思潭 (Sadam，思想者池塘) 也不过是给潺潺的流水平添了几分景色；暝庭 (Myeongjeong，冥想花园) 让我们从目不暇接的美景之中得到片刻休息，之后方可继续欣赏美景；卧寺 (Wasa，爬行寺) 坐落在一个偏远的山谷之中，隐藏在水流之后；瞻坛 (Cheomdan，天文观测台) 吸引着人们向上攀爬，仰望群星——为了让人能够理解这些建筑的理念，我对这些名字也着重进行了设计。

除此之外，我还给植物园设计了其他辅助设施，例如：前门、生态卫生间、观景台、长椅和灯具，这些设施分布在植物园的各处，整体上构成了一种连贯性。因为刘宰成主席特别喜欢看表演，所以我在植物园内，按照不同的环境特点设计了不同的小舞台。

随后，刘宰成主席开始亲自给各个设施和场所命名——致虚门 (Chiheomun)、别有洞天 (Byeolyudongchen)、悟塘 (Odang)、风屑几千年 (Pungseolkicheon-yeon)、小白洗心台 (Sobekseshimdae)、啸岗弹琴台 (Sogangtanggeumdae)、平田 (Pyungjeon)——他将汉字如诗般组合在一起，尽管我平时经常学习和练习中文，还是让如此高水平的汉字组合难住了。正是名字赋予了万物生机。

但同样，也有部分设计我没有参与，刘宰成主席将这些任务交给了我同样喜欢的建筑师——阿尔瓦罗·西扎，他的设计包括一所小教堂，一个拥有精致空间感的多功能表演场地以及一个观景台，将西扎独特高雅的建筑感展现得淋漓尽致。有两座建筑看起来像是韩屋 (韩式建筑)，让我觉得有点疑惑，但我毫不怀疑，这两座建筑会随着时间的推移而融入到植物园的自然环境之中。

随着植物园的建造接近尾声，我接到了自己最后的任务：位于植物园入口的住宿设施。这里的美不是只看一眼就能领略到的，尽管我来过了无数次，但这里的景色每次都滋养着我贫瘠的思想，治愈我浑浊的情感。古老的榠楂、紫薇、光叶榉树、松树和野花在一年四季、从清晨到日落，呈现出完全不同的景象。我曾在夜深人静的时候，沐浴在繁星之下；也曾在日出之时，透过迷雾，欣赏群山的轮廓，让我感受到自己的渺小。

这让我意识到，为了饱览这里的自然景色，住宿设施是很有必要的。但刘宰成主席似乎不太愿意去管理一家传统的酒店，因此他最初并不愿意接受这个想法。但是最终，我们达成一致，同意将住宿设施变成"Gochimjeongsa"，意思是"单独居住的私密卧室"，目前我们正处在设计的最后阶段。

挪威建筑师克里斯蒂安·诺贝格-舒尔茨写了《地方守护神：迈向建筑现象学》一书，他在书中写道："建筑的基本行为是要去理解一方土地的'使命'。"我明白植物园脚下土地的意义了吗？我试过去理解，但可能还不够，又或者我甚至背叛了这片土地的"使命"。但我从未怀疑过任何一片土地所拥有的活力。无论多么伤痕累累，土地都能治愈自己，战胜苦难，甚至把自己的苦难用优美的文字传递给下一代。土地耐心地等待着，仿佛在告诉我们，过往已离我们远去。因此，正如荷尔德林告诉我们的那样，人之居也如诗。

I cannot start the story without talking about Chairman Yoo Jae-sung of Daegu. Yoo, the owner of Taechang Steel, is an influential yet reserved sponsor and patron of the artistic and cultural development of the Daegu-Gyeongbuk Region in South Korea.

Several decades ago, Yoo once prevented an entire set of quince trees hundreds of years old from being smuggled to overseas by purchasing them for a generous fee. After this incident, he used to be asked to collect rare quince trees from all over the country and thus decided to start an arboretum. He came to acquire a plot of land in Bugye of Gunwi-gun, Gyeongsang-bukdo upon this decision.

1. 暝庭
2. 瞻坛
3. 思潭
4. 卧寺
5. 玄庵
6. Gochimjeongsa (单独居住的私密卧室)
7. 致虚门

1. Myeongjeong
2. Cheomdan
3. Sadam
4. Wasa
5. Hyeonam
6. Gochimjeongsa
7. Chiheomun

It was 10 years ago when I designed the annex of his home of 40 years and even gave it the name "Moheon", that I formed a personal relationship with him. Moheon meaning "somebody's house" is a fitting name for Yoo who prefers to keep a low profile. Perhaps, for this reason, Yoo discussed his arboretum with me beginning from its concept. I named the arboretum "Sayuwon", perhaps a rather direct name, suggesting that it be a place for reflection and meditation. "Bugye Arboretum Sayuwon". Confucius once said, "The beginning of wisdom is to call things by their proper name," and the name Sayuwon gave us a clear idea of the character of the arboretum. Beyond the arboretum, the building facilities had to offer a reflective place for us to engage in deep introspection.

Above all, my first encounter with the land was so intense that it gave me the confidence that the land necessitates such a character. The site, consisting of two ravines and three ridges and situated between Mt. Palgong to the south and Mt. Dobong to the north, did not have anything artificial in view when I climbed to the highest point on my first visit. Ah! The land was truly autonomous, situated deep into nature to feel as if it was only me who existed between earth and sky. Today, the site totals around 1,000,000m², but visually connected to adjacent mountains, and the land feels far larger reaching a hundred million square meters.

Jung Young-sun, a prominent Korean landscape architect whom I have great faith in, shaped the overall framework of the masterplan, subdividing the land graciously. Kawagishi Matsunobu, a Japanese landscape architect, who expertly designed and executed the landscape adjacent to Yoo's house in Daegu, providing valuable insight from time to time. Together with their help, the arboretum's distinctive landscape design began to take shape.

Sunshine reaches the depth of the forest by carefully managing its density, giving life to wildflowers and grass, while also inviting birds with a new stream of water. Transplanted quince trees lull their immediate surroundings standing firmly against the beautiful backdrop of crape myrtle trees. Pine trees, each astronomical in value, find their place in the forest to show off their beauty perhaps as the most beautiful of all trees. The new landscape is surely a sublime ensemble.

This land requires the landscape and the trees to become the focal point, and the architecture not to have an extravagant form. Architecturally, the building remains modest, merely providing a good place to observe the landscape through thoughtfully positioning the building within the site. Thus, "Hyeonam (Black Cottage)", my first building here, as well as the other buildings in the site are either sitting low close to the ground or buried into it. My only wish was to create architecture as if it had always been there.

Hyeonam, a space between Earth and Heaven; Sadam (Thinker's Pond), a mere backdrop to the performances of burbling water; Myeongjeong (Meditation Garden), a brief breakpoint from the beautiful landscape scenery for us to enjoy it again; Wasa (Crawling Monastery), a shelter by the water in a remote valley; Cheomdan (Astronomical Observatory), a seductive climb to gaze at stars – I styled these names to clarify their concepts.

Beyond this, I designed other auxiliary facilities for the arboretums – like the front gate, ecological toilet, the observatory, long benches, and light fixtures – scattered across the site to form an overall coherence in their language. Small performance stages are built in accordance with the particularity of their immediate surroundings within the arboretum for Yoo who particularly enjoys performances.

Then, Mr. Yoo gave names to facilities and places himself – Chiheomun, Byeolyudongchen, Odang, Pungseolkicheon-yeon, Sobekseshimdae, Sogangtanggeumdae, Pyungjeon – poetically combining Chinese characters so high-level that even I

have difficulties despite my regular study and practice. Indeed, names give vitality to things.

There are also several parts of the project that I was not involved in designing. Yoo commissioned Alvaro Siza, an architect that I am also fond of, for a small chapel, a multi-purpose performance venue with an exquisite sense of space, and an observatory that reveals Siza's unique and elegant architectural sensibility. There are also two buildings appearing as a Hanok (Korean architecture) that I find questionable; but, I have no doubt that, over time, these two will be enmeshed with the natural surrounding of the arboretum.

As the construction nears its completion, I was commissioned for my final project: an accommodation facility by the entrance to the arboretum. The beauty of the arboretum cannot be absorbed in a single glance. Despite the countless visits to the arboretum, its scenery tends my barren thoughts and heals my murky sensibility each time. Centuries-old quince, crape myrtle, zelkova, pine trees, and wildflowers present themselves completely differently between spring and summer, fall and winter, and from morning to evening sunset. In the middle of the night in utter darkness, I once stood showered by starlight, while I have also felt powerless watching the silhouette of the mountains peaking through a deep layer of fog during sunrise. Subsequently, an accommodation facility was an absolute necessity to gain a thorough experience of its natural scenery. This was an idea Yoo was initially reluctant to commit to, given his reluctance in managing a conventional hotel. But, at last, we agreed to see the accommodation facility as a retreat house called "Gochimjeongsa", meaning "an intimate bedsit for solitary habitation", which we are currently in the final stage of designing.

Christian Norberg-Schulz, who wrote *Genius Loci: Towards a Phenomenology of Architecture*, says, "the basic act of architecture is to understand the 'vocation' of the place." Did I understand the vocation of the place that the arboretum occupies? I have tried but it was likely not enough, or perhaps I even betrayed its vocation. But I have never doubted the vitality of a place. No matter how torn and scarred, lands have the capability to heal itself, to overcome it, and to pass on even their sufferings through beautiful words to the next generation. Lands patiently await as if to say that the past is already before us, and thus, as Holderlin tells us, poetically man dwells.

Seung H-Sang

玄庵
Hyeonam

在大邱市一所名为"某轩"(Moheon)的小房子中,我们的客户探索出了新的生活乐趣,他开始在韩国军威郡(Gunwi) 100ha的山区之中建造植物园。在移植了具有600年历史的温梓树之后,森林又因砍伐围垦而变得稀疏。他观察到了自然的变化,因而想要在那里建造第一所用来居住的房子。房子的选址位于一座山之中,在这里只能看到脚下崎岖的地形和头顶的天空。房子朝西,我们能够看到夕阳从水库上方缓缓下沉,因此在冬日里,夕阳西下时会给屋子带来融融暖意。

客户希望他未来的植物园能成为一个冥想之处。这就是为什么在入口沿路建造了五个不大的人工花园。穿过茂密的松树林,便会来到这个小屋。虽然这些花园都很小,但每个花园周围的斜坡和平地都形成了鲜明的对照,触发人们的冥想。在花园的尽头,一座耐候钢打造的建筑坐落在一条笔直的斜坡之上。再往上走是一座假山,在此处可以看到下面的山谷。穿过大门,进入到这座耐候钢建筑之中,眼前无与伦比的自然景观将人们团团围绕。这里没有什么建筑理念,只有人和自然存在于此,万籁俱寂。

若是游客有幸在日落时来到这里,他们就可以欣赏到美丽的落日余晖。爬上小山,坐在泛着银色光泽的草地中央凉爽的耐候钢椅子上时,便会觉得自己与自然融为了一体。这是一个极其孤独的时刻,也是一个适合思考哲学的时刻。这就是为什么这所房子被取名为玄庵(Hyeonam),字面意思是一个黑色、阴暗的小屋。

当人们从远处看到这所房子时,会发现它从地面凸出,像是经历了漫长的等待之后刚刚破土而出来到世上一样。这就是这所房子还没有完工却仍然存在于此的原因。

The client, who discovered a new joy of life in a small house named Moheon in the City of Daegu, began constructing an arboretum in a 1,000,000m² mountain area in Gunwi, Korea. As he transplanted 600-year-old quince trees, thinned the forest, and observed the change of the nature, he wanted to build the first house to live there. The site is located in the middle of the mountain where people can see nothing but the rugged terrain and the sky. It faces the west, giving us a chance to see the sun going down over the reservoir and giving warmth in the winter season along the path of the sun.

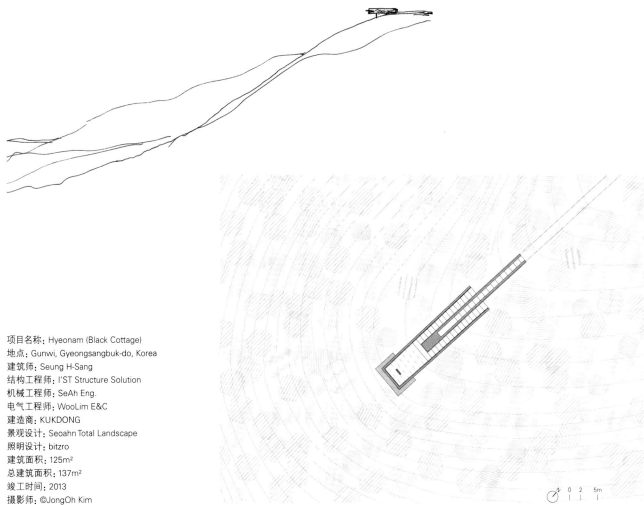

项目名称：Hyeonam (Black Cottage)
地点：Gunwi, Gyeongsangbuk-do, Korea
建筑师：Seung H-Sang
结构工程师：I'ST Structure Solution
机械工程师：SeAh Eng.
电气工程师：WooLim E&C
建造商：KUKDONG
景观设计：Seoahn Total Landscape
照明设计：bitzro
建筑面积：125m²
总建筑面积：137m²
竣工时间：2013
摄影师：©JongOh Kim

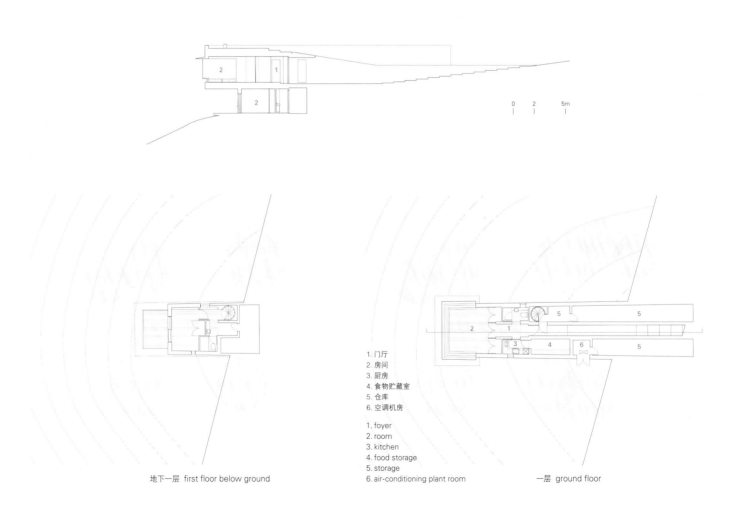

1. 门厅
2. 房间
3. 厨房
4. 食物贮藏室
5. 仓库
6. 空调机房

1. foyer
2. room
3. kitchen
4. food storage
5. storage
6. air-conditioning plant room

地下一层 first floor below ground

一层 ground floor

The client hoped that his future arboretum would be a place for meditation. This is why the five small artificial gardens along the access road were made, passing through the dense pine tree forest, to this cottage. While these are all small, it seems that the contrast of the slope and the flatland around each of the gardens would be a trigger for meditation. At the end of the gardens, people see a Corten steel structure along the straight slope. Above this structure, we encounter an artificial hill while seeing the valley below. Entering the structure through the door, people find themselves overwhelmed by the deep embrace of the majestic natural landscape. No architectural concept exists there, only he or she and nature

remain, in complete silence.

If visitors are fortunate enough to be there at sunset, they could see the beautiful glow of the setting sun. When they climb on the hill and sit down on a cold Corten steel chair in the middle of silver grass, they will feel like they are a part of the nature. This is a completely solitary moment, and a moment for philosophy. This is why the house was named Hyeonam, which literally means a black and shady cottage. When people see this house at a distance, they will find it protruded from the ground as if it just came out of the underground into the world after waiting for a long time. That is why this house is yet to be completed and remains still.

思潭
Sadam

我们利用池塘、舞台和舞台背景在水边搭建了许多供人欣赏文化表演的设施。舞台的背景是一堵由耐候钢制成的墙。这堵墙不会腐蚀,笔直地矗立在那里,就像是被人从山里拉出来的一样,成为一场精彩演出的背景。在墙的后面有一家小餐厅,以及附属的表演设施,游客可以在没有表演的时候,在这里欣赏海滨的景色。

舞台与屋顶相连,陡峭的楼梯连接到耐候钢墙的一端。屋顶上还有一个舞台,如果有需要,可以在不同的楼层进行表演。当然,屋顶的舞台也可以作为观景台,用于眺望周围的景色。尤其是当太阳下山时,观众坐在池塘对面斜坡的长凳上,观看池塘里倒映出的表演,堪称一幅美丽的画卷。

Facilities for enjoying cultural performances at the waterfront have been established by creating a pond, a stage, and its background. A wall made of Corten steel, which does not corrode, is the background of the stage, and stands up as if it were pulled out of the hill as the backdrop for a beautiful performance. Behind the wall there is a small restaurant, as well as subsidiary facilities for the performance where visitors can enjoy the scenery of the waterfront when there is no performance.

The stage is linked to the rooftop with a steep stairway connecting to one end of the Corten steel wall. There is another stage on the rooftop which makes a performance on different levels possible if needed. Of course, the stage on the rooftop can function as an observation deck for watching the surroundings. Especially when the sun sets, the scene for the audience, sitting on long benches on the slope opposite the pond and watching the performance reflected on the pond, looks like a beautiful picture.

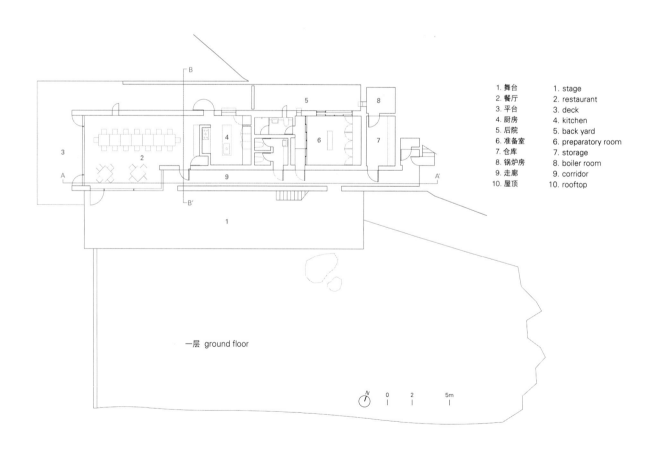

1. 舞台   1. stage
2. 餐厅   2. restaurant
3. 平台   3. deck
4. 厨房   4. kitchen
5. 后院   5. back yard
6. 准备室   6. preparatory room
7. 仓库   7. storage
8. 锅炉房   8. boiler room
9. 走廊   9. corridor
10. 屋顶   10. rooftop

一层 ground floor

项目名称:Sadam (Thinker's Pond) / 地点:Gunwi, Gyeongsangbuk-do, Korea / 建筑师:Seung H-Sang / 结构工程师:The Naeun Structural Eng. / 机械工程师:DE-Tech. 电气工程师:WooLim E&C / 建造商:EONE / 景观设计:Seoahn Total Landscape / 照明设计:bitzro / 用地面积:137m² / 总建筑面积:128m² / 竣工时间:2016 摄影师:©JongOh Kim

南立面 south elevation

A-A' 剖面图 section A-A'

B-B' 剖面图 section B-B'

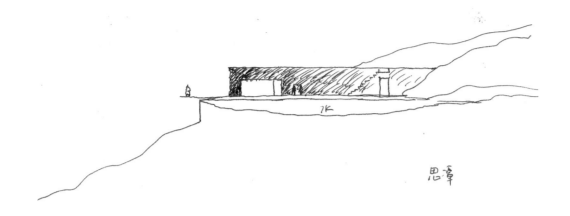

思潭

瞻坛
Cheomdan

整个场地的供水由两个大水箱负责，其中一个建在地面上。地面上的这个水箱可以作为另外一个观景台，从西面俯瞰植物园。这个用混凝土制成的水箱有一个沿着墙壁向上的楼梯。水箱分成几个部分，看起来像一个小堡垒。水箱也可以作为一个观测台，用来仰望镌刻在天象图上的星星。这就是这座建筑被取名为瞻坛（Cheomdan）或者天文观测台的原因。周边区域自成一派，勾勒出一种独特的景色，也让石头砌成的天文台的小门和一望无尽的草地交相辉映。

One of the two large water tanks supplying the entire site is built on the ground, and it may be used as another viewing platform to look over the arboretum from the west. This concrete structure has a staircase that rises up along the wall, and the structure is divided into several segments that look like a small fort. It also serves as an observatory to look at the stars engraved on the sky chart. This is why the name of this structure is called Cheomdan or astronomical observatory. The vicinity of this area forms a special landscape, matching the small doors of the astronomical observatory made of stone with the far-reaching meadow.

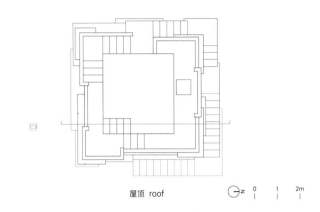

屋顶 roof

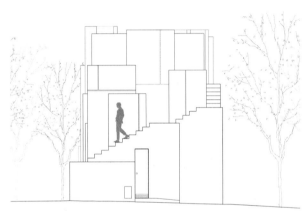

西立面 west elevation

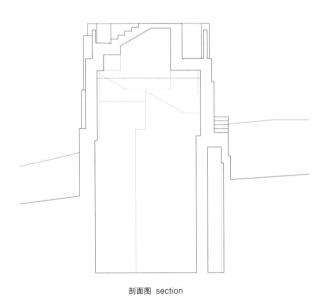

剖面图 section

项目名称：Cheomdan (Astronomical Observatory)
地点：Gunwi, Gyeongsangbuk-do, South Korea
建筑师：Seung H-Sang
结构工程师：The Naeun Structural Eng.
机械工程师：DE-Tech.
电气工程师：WooLim E&C
建造商：YoungJo Construction Co.
照明设计：NEWLITE
建筑面积：35m²
总建筑面积：35m²
竣工时间：2019
摄影师：©JongOh Kim

暝庭
Myeongjeong

按原先的设计规划，暝庭本来是植物园北峰的观景台，现在却成了游客逛遍植物园几乎所有美景之后会来到的地方。为了让植物园的景色在游客的记忆中更加多彩，就必须让他们记住这些美丽的景色。因此，暝庭不仅仅是一个观景台，还是一个让游客能够进行冥想和反思的地方。首先，这里被取名为暝庭（沉思园），全部设施都建在地下。在地面上只能看见混凝土制成的影壁，但即使是这堵墙也被隐藏在了杉林之中。长长的墙壁通向一个斜坡，上坡之后，你会看到一个小洞口和一条长通道。沿着这条狭窄的通道直行，然后转弯，便会看到一个下行的楼梯，继续向下走27m，然后转弯，眼前的空间绝对出乎你的意料。水沿着石墙流出，流满庭院的地面。流满水的地面上有一条通路，沿着通路可以走到对面的舞台。这里没有树，只有一条长长的混凝土长椅在水流前方等待着你。

穿过流水的石墙，便来到一个跨度很长的静默空间，这里有像小庇护所一样的帐幕，只能听到水从石墙上流下来的声音。但寂静还是更胜一筹，因此你在这里唯一能做的事情就是"冥想"。

迈上两堵石墙之间狭窄而陡峭的楼梯，向南延伸的植物园与柔和而精致的八公山剪影便映入眼帘，仿佛将整个世界收入囊中，何其壮观！这里的风景与你第一次看到的有所不同，成为一个等待你再次探索的新世界。

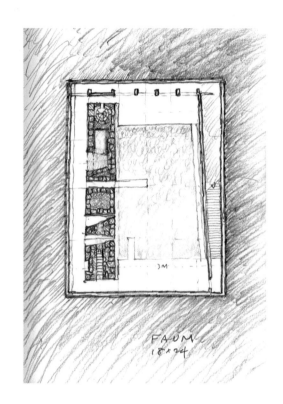

This place, planned as an observation deck on the northern peak of the arboretum, is the place where visitors come after seeing almost all the beautiful scenery of the arboretum. In order to make the view of the arboretum more beautiful to those who bring all the memories of the arboretum, it is necessary to keep the memory in mind. That is why this place is not a mere observation deck, but a place for the visitors to reflect on themselves. First of all, this place was named Myeongjeong (Meditation Garden), and the entire facility was pushed into the ground. Only the concrete front wall is visible, but even the concrete wall is concealed by a fir forest. When you go up a slope led by a long wall, you will see a small opening and a long aisle. If you go along this narrow passage and turn, you will find descending stairs. When you go 27 meters forward down the stairs and turn, you will encounter an unexpected space. There, you will see water running down the stone wall to cover the floor of a courtyard, and a stage on the opposite side connected by a pathway over the water. A long bench of concrete awaits you in front of the water. No trees can be seen here.

When you are past the wall of flowing water, a long silent space appears, and there are tabernacles like small sanctuaries in this space. You can hear only the sound of the water running down the stone wall, but the sound of silence prevails. What lies here is only meditation.

It is truly spectacular to see the landscape of the arboretum spreading to the south and the soft and delicate silhouette of Mount Palgong, which seems to have enveloped the whole world when you go up the narrow and steep staircase made between the stone walls. This landscape is different from the one you saw first and it will be a new world again.

南立面 south elevation

1. 入口走廊
2. 冥想庭院
3. 观景台

1. access corridor
2. meditation court
3. observatory

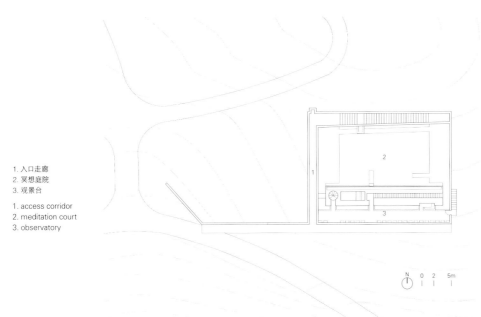

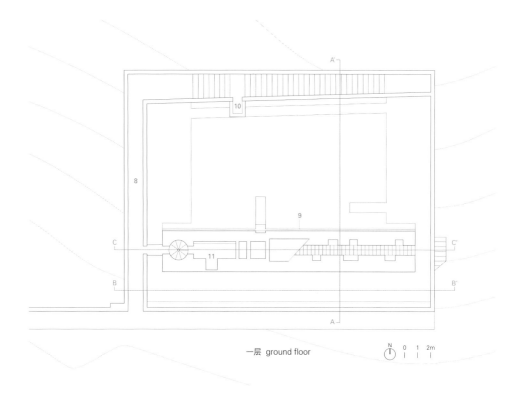

一层 ground floor

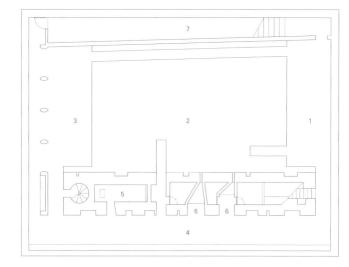

1. 平台
2. 流水庭院
3. 舞台
4. 冥想回廊
5. 冥想室
6. 祭具室
7. 冥想者之屋
8. 走廊
9. 小瀑布
10. 阳台
11. 纪念室

1. platform
2. water court
3. stage
4. meditation corridor
5. meditation room
6. sacristy
7. thinker's room
8. corridor
9. cascade
10. balcony
11. memorial room

地下一层 first floor below ground

1. 观景台
2. 冥想回廊
3. 小瀑布
4. 流水庭院

1. observatory
2. meditation corridor
3. cascade
4. water court

A-A' 剖面图 section A-A'

1. 走廊
2. 观景台
3. 冥想回廊

1. corridor
2. observatory
3. meditation corridor

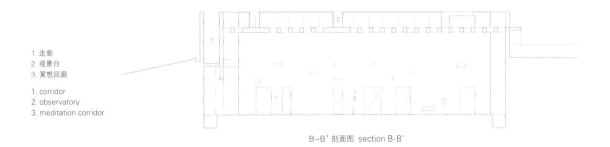

B-B' 剖面图 section B-B'

1. 走廊
2. 纪念室
3. 冥想回廊
4. 冥想室
5. 祭具室

1. corridor
2. memorial room
3. meditation corridor
4. meditation room
5. sacristy

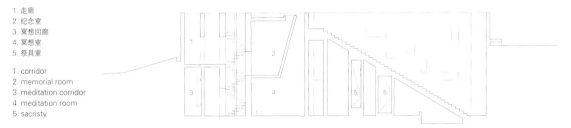

C-C' 剖面图 section C-C'

项目名称：Myeongjeong (Meditation Garden) / 地点：Gunwi, Gyeongsangbuk-do, South Korea / 建筑师：Seung H-Sang / 结构工程师：The Naeun Structural Eng. / 机械工程师：DE-Tech. / 电气工程师：WooLim E&C / 建造商：YoungJo Construction Co. / 照明设计：NEWLITE / 建筑面积：150m² / 总建筑面积：150m² / 竣工时间：2019 / 摄影师：©JongOh Kim

卧寺
Wasa

植物园下面陡峭的山谷里建有三个池塘。尽管这三个池塘大小和深度各不相同，却是彼此相连的。这三个池塘的设计理念是：游客需要在池塘边休息。但是，池塘位于整个植物园中最偏远的地方，因此为了让这里值得一看，就需要一座具有特色的建筑。这座建筑便是一座寺庙。但是，在此位置上由于法律规定，我们无法盖一幢可居住的建筑。因此，我们决定盖一座充满艺术气息的建筑，内部拥有广阔的空间，这样，就不需要为其配备必需的生活设施了，如保温层和防水层。因此，我们可以专注于这座建筑的本质：空间。

这座充满艺术感的建筑应该沿着那三个彼此连接的池塘而建，这一概念催生出了这样一个想法：我们可以盖一座呈俯卧状的寺庙。这里的结构像是一个跨度很长、用耐候钢制作的盒子，其内部空间按我们设想的一样，可以分为教堂、餐馆、图书馆和宿舍。整体看上去就像一个富有生命的有机体，其高度可以随着地形的变化而变化。建筑一侧的塔楼是卫生间。这座建筑中悬挂着小风铃，游客被钟声唤醒，感觉自己仿佛置身于山中的寺庙。如果愿意花点时间停下脚步，靠在这里，看看池塘里的景色，便会感受到大千世界的美丽。

Three ponds were built in the steep valley down the lower part of the arboretum. Although they are different in height and size, they are connected with each other. It was considered that visitors would need resting shelters along the ponds. It was, however, the most remote place in the entire arboretum and needed a special facility worth visiting. It was a monastery. However, due to legal requirements, it was not possible to construct a habitable building. Instead, it was decided to build an artistic structure with space, for which there was no need to equip the structure with measures indispensable for living such as insulation and waterproofing. It was therefore possible to concentrate on the essence of the space of the structure.

The artistic structure should be long to run along the three connected ponds. This concept suggested an idea of a monastery lying down. As such, the interior space of the long Corten steel box was divided for an imagined arrangement of such places as a chapel, restaurant, library, and dormitory. The facility is made to look like an organism, with its height changing according to the terrain. The tower on the side of the structure is the bathroom, where small wind-chimes are hung so that visitors can feel that they are in a temple, evoked by the bell sound, in the middle of the mountains. If you take some time to stop and lean on this facility just for a while and look at the water in the pond, you will feel how beautiful the world is.

1. 祭坛 2. 教堂 3. 食堂 4. 牧师会礼堂 5. 钟楼/卫生间
1. altar 2. chapel 3. refectory 4. chapter house 5. bell tower/toilet
一层 ground floor

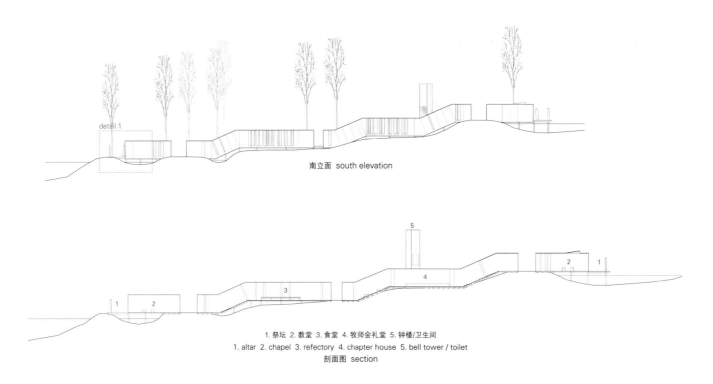

南立面 south elevation

1. 祭坛 2. 教堂 3. 食堂 4. 牧师会礼堂 5. 钟楼/卫生间
1. altar 2. chapel 3. refectory 4. chapter house 5. bell tower / toilet
剖面图 section

项目名称：Wasa (Crawling Monastery) / 地点：Gunwi, Gyeongsangbuk-do, South Korea / 建筑师：Seung H-Sang / 结构工程师：The Naeun Structural Eng.
建造商：YoungJo Construction Co. / 照明设计：NEWLITE / 建筑面积：130m² / 总建筑面积：130m² / 竣工时间：2019 / 摄影师：©JongOh Kim

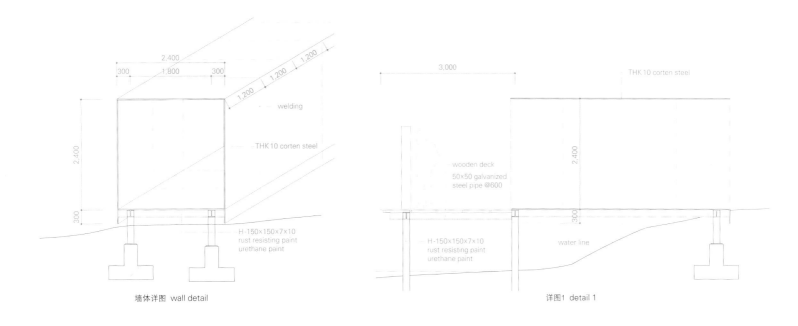

墙体详图 wall detail   详图1 detail 1

鸟鸣寺（鸟巢）
Birds' Monastery Birds' Nest

鸟鸣寺的设计雏形是一个非军事区的装置艺术项目，并曾经在几家美术馆展出，包括日本的原美术馆。由于水是储存起来并流经整个植物园的，鸟儿成群结队地栖息在这里，因此鸟儿们需要一个"家"，而这个尚未完成的项目便有机会实现这一点。根据鸟儿不同的飞行高度，鸟鸣寺的设计分为五层，有着寺庙的寓意。人们可以偶尔在此处逗留。竹子是主要的建筑材料。竹子和周围的环境打成一片，但却总有一天会倒塌，成为大自然中的过客。

This facility was originally designed as a DMZ installation art projects and exhibited in several galleries, including the Hara Museum in Japan. As water is stored and flows out over the whole arboretum, a shelter for birds flocking in is required and the unbuilt project has a chance to be realized. The design, with five stories according to different flying heights of birds, is a metaphor of a monastery. Occasionally people could stay here briefly. Bamboo is the main material. It connects well with the environment and someday it should collapse and remain merely as a memory in the natural environment.

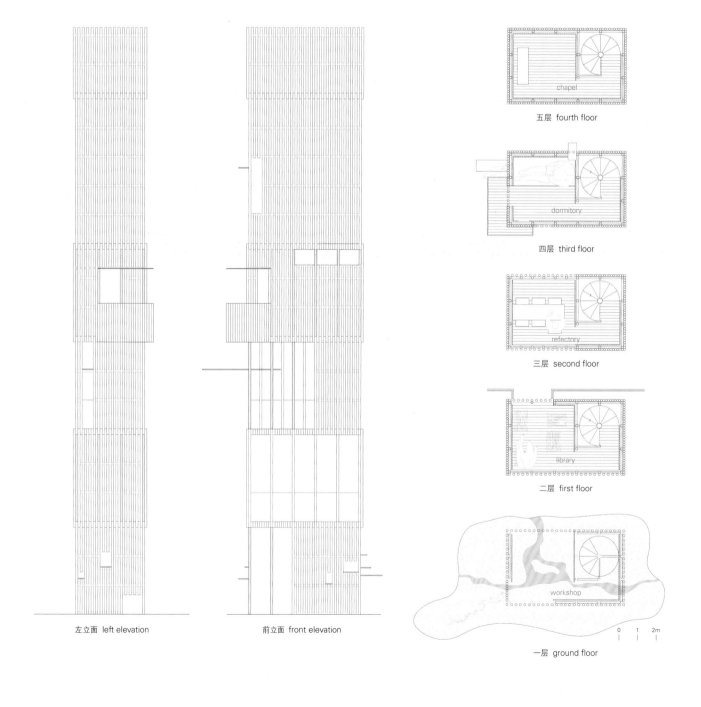

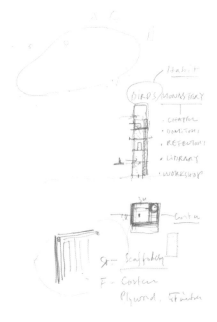

Gochimjeongsa（精品酒店）
Gochimjeongsa Boutique Hotel

除了白天以外，植物园的夜景也值得一看，尤其是在日出和日落之时的景色更是无与伦比地美丽。因此，酒店成了植物园必不可少的设施。酒店建在入口区域，内有50间客房，其设计理念也符合"冥想"的初衷。每间客房都像僧侣的房间一样，小巧而简朴。每间客房的每个角落都与美丽的大自然交融在一起。尽管酒店"穿着"坚硬的混凝土外衣，但温暖的室内和郁郁葱葱的树木足以给那些想待在外面与想要独处的人带来慰藉。

This arboretum has fantastic beautiful scenery, not only in the daytime but in the night, and especially at dawn and sunset. A hotel will be an essential facility. With 50 rooms and built in the entrance area, it should be designed as meditative. Every room is like a monk cell, tiny and simple. But every room and every space in this monasterial hotel is in contact with beautiful nature. It is enclosed by the tough texture of concrete, but its warm interior and plentiful trees could comfort those who want to stay out and stay alone.

东立面 east elevation

西立面 west elevation

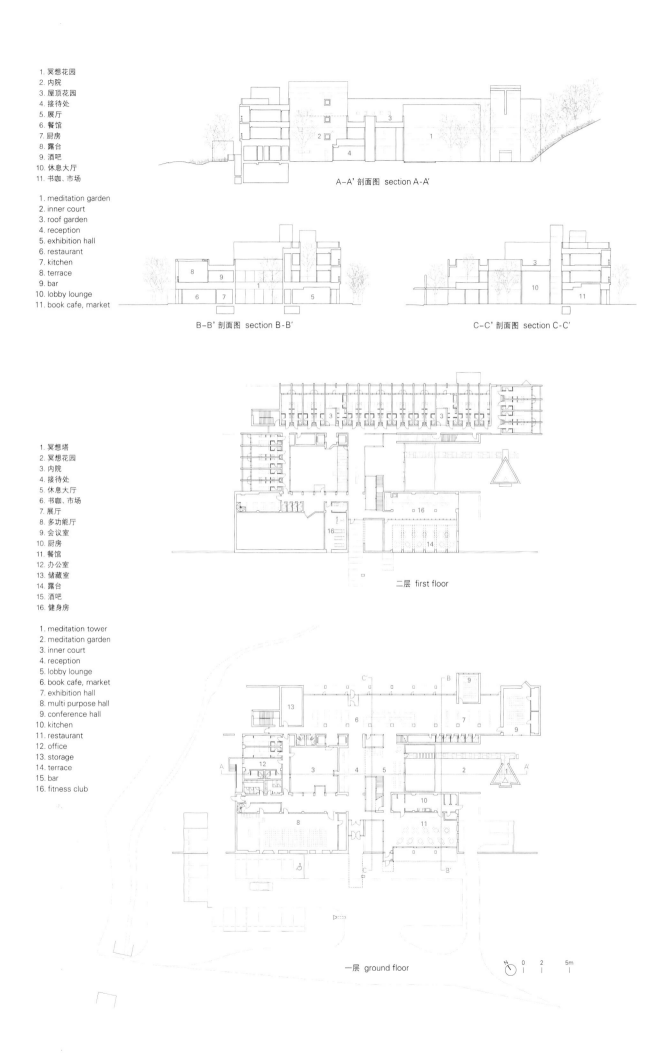

配套设施
Supporting Facilities

　　植物园内还有一些配套设施，例如：入口处的停车场、长椅、沿漫步小巷搭建的临时舞台、路灯、室外楼梯、一个观景台，以及园内随处可见的生态厕所，这些设施使用了耐候钢、木头或混凝土作为建筑材料，在表现植物园的形象上发挥着十分重要的作用。有时，这些配套设施具有纪念意义的外观也强化了植物园的场所意识。

There are several supporting facilities in this arboretum. The parking lot around the entrance, the benches and makeshift performance stages along the strolling alleys, a lighting pole, outside stairs, a viewing platform and the ecological restrooms scattered over the place, are all made of Corten steel, wood or concrete, which plays a big role in expressing the identity of the whole arboretum. Their sometimes monumental look enhances the sense of the place.

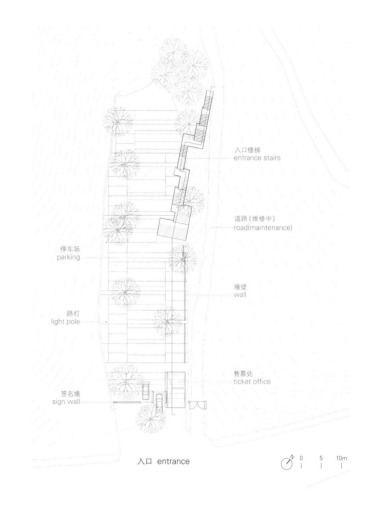

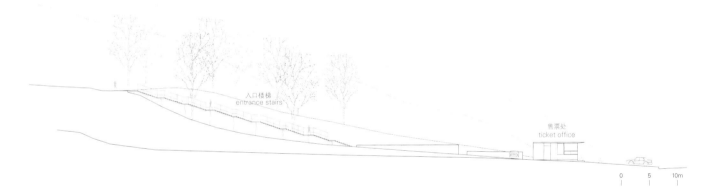

入口楼梯
entrance stairs

售票处
ticket office

0 5 10m

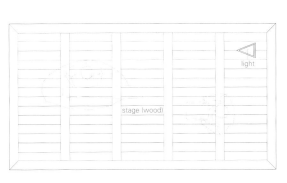
室外舞台B一层 outdoor stage-B, ground floor

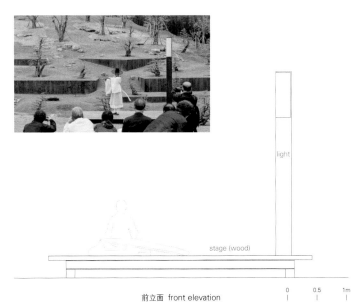
前立面 front elevation

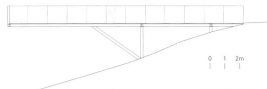

观景台B剖面图
observatory-B, section

观景A一层
observatory-A, ground floor

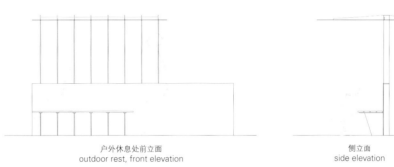

户外休息处前立面
outdoor rest, front elevation

侧立面
side elevation

独与群

思惟园与承孝相的建筑

Alone and Together

Sayuwon and the Architecture of Seung H-Sang

Hyungmin Pai

正如人固有一死,建筑也终将坍塌,灰飞烟灭。无论建筑为了扬其赞助者的名,建得多坚固,所有建筑都难敌万有引力。只有存在过的记忆才是不朽的。

——承孝相《论黄龙寺的废墟》,2006年。

在当今时代,人类开始认识到,自身文明在地球上是如此渺小,而且很可能已经临近其短暂寿命的终点。在人类的所有手段中,建筑在传统上扮演着衡量时间、衡量人类抱负的角色。为了延长建筑在地球上存在的时间,我们改变了建筑,这就是我们通常所说的,为"可持续性"所做出的努力。正如我们解决了这样一个急迫的挑战,建筑作为时间尺度的作用也受到了质疑。面对诸多挑战,思惟园让我们能够去面对时间的本质。思惟园是一座私人植物园,位于朝鲜半岛东南部军威郡占地100ha的山区之中。建造植物园的这一想法最早由Taechang钢铁公司的主席刘宰成提出,其实在此前这个想法就已经酝酿了很久了。刘宰成白手起家,不走寻常路,热爱资助艺术,他一直以来的梦想就是拥有一个植物园。用他自己的话来说,他相信:"钢铁为生计,花园冶情操"。在20世纪80年代末到90年代初,刘宰成买下了许多榅桲树和鹅耳枥树,为了保护这些树,就需要给这些树一个永久的"家"。这时,刘宰成梦想中的植物园就已经初具规模了。随着植物园度过了初创期,刘宰成开始寻找建筑师来设计第一批建筑,让这里一望无际的景观不再那么空旷。

自2012年的春天以来,承孝相设计了一些建筑:玄庵(Hyeonam)是一个小观景房,也是他在这里设计的第一座建筑;思潭(Sadam),这里有一家自助餐厅和海滨表演舞台;暝庭(Myeongjeong)是一个引人深思的户外流水庭院;卧寺(Wasa)是一座由耐候钢建成的建筑,坐落在一个沿着山谷溪流建造的人工池塘上方;瞻坛(Cheomdan)是一座小眺望塔,内有一个水箱;停车场;一些户外厕所;最后,还有一家酒店,这是植物园最大的建筑,也是唯一由承孝相设计却还没有开工的建筑。Jung Young-sun与Kawagishi Matsunobu设计了这里的自然景观;阿尔瓦罗·西扎设计了一座家庭教堂、一座眺望塔、一个画廊;书法

As man cannot deny the fatality of death, architecture ultimately falls down and disappears. No matter how firmly it is built to celebrate the glory of its patron, there is no building that can finally resist the law of gravity. What remains is only memory.
– Seung H-Sang, "On the Ruin of Hwangryong Temple," 2006.

We live in an age where humanity is beginning to recognize that its civilization may be at the tail-end of its very short life-span on earth. Among its devices, architecture has traditionally played the role of a gauge of time, of a measure of humanity's aspirations. As much as we address the urgent challenge of changing architecture toward prolonging its existence on earth, what we often call the efforts toward "sustainability," the role of architecture as a temporal gauge is also called into question. Among the many questions that it poses, Sayuwon is a place that confronts us with the nature of time. Arboretum Sayuwon is a private park, a 1,000,000m² mountainous area of Gunwi in the south-east region of the Korean Peninsula. Initiated in 2006 by Yoo Jae-sung, Chairman of Taechang Steel, it has a long and personal pre-history. Yoo is a self-made man and a maverick, a patron of the arts with a life-long dream of creating an arboretum park. To use his own words, he is someone who believes that while "steel is a part of my work, the garden is part of my life". This dream garden began to take concrete shape in the late 1980s and early 1990s when the quince trees and carpinus trees that he had acquired and protected needed a permanent home. As the arboretum passed its years of infancy, Yoo began to seek architects to design the first structures to occupy its vast landscape.

Since the spring of 2012, Seung H-Sang has designed several structures: Hyeonam, a small viewing house, the first of his designs to be built; Sadam, a cafeteria and waterfront performance stage; Myeongjeong, a contemplative outdoor water courtyard; Wasa, a corten that sits over an artificial pond created along the valley stream; Cheomdan, a small lookout tower that encases a working water tank; the parking lot; series of outdoor toilets; and finally, a hotel, the largest facility in the park and the only structure of his design that has not been built. They are structures that co-exist with the landscape designs of Jung Young-sun and Kawagishi Matsunobu; a family chapel, a look-out tower, and a gallery space designed by Alvaro Siza; the calligraphy of Wei Eerang;

家Wei Eerang设计了这里的书法;客户(刘宰成)自己也加入了很多想法和意见;承孝相设计的建筑和以上种种都能完美兼容。在植物园设计的第一座建筑建在山脊上,自那以来,已经过去十年了。十年来,思惟园仍然是一个开放的、尚未竣工的项目,是一个庞大的植物园,但并没有总体的规划。唯一可能做出总体规划的人便是客户本人,他大胆的举动往往不符合任何逻辑和计算。那么思惟园到底是什么?是一场商业冒险还是个人的梦想?思惟园的体验维度怎样?思惟园的实际边界有多大?植物园里的建筑会有实用性吗,还是成了无用建筑中的"稀有品种"?但可以确定的是,思惟园体现的是人工自然:是一个美丽的谜题,让所有身处其中的人都为之疯狂,但又无法抗拒。即便没有总体规划,思惟园仍然具有一个认知和感知的体系,这一体系是其内部许多因素相互作用而形成的。

承孝相最突出的能力在思惟园的设计上体现得淋漓尽致而又十分微妙。显然,作为一名建筑师,他有许多作品,但他最重要的角色却是"风水师",即客户的顾问,负责解释建筑类型、建筑地点和建筑方式。玄庵是思惟园的第一个建筑项目,是一座没有特意去选址建造的小房子。承孝相先是找到了建筑场地,创造出了一个小空间,在这样的小空间里,将思惟园浩瀚的山脉放大成了一幅全景。他的第二个项目是思潭。思潭要同时作为咖啡馆和户外剧院。因此,承孝相使用现有的池塘将观众和舞台分隔开来,同时还设有标志性建筑,提示游客已经进入植物园的中心。随后,客户要求承孝相设计一座永久性的眺望塔,来取代植物园最高处的那个临时结构物。但和刘宰成期待的直上直下的建筑物不同,承孝相另辟蹊径,设计了现在的暝庭——这是一个陷进地下的广场,身处广场,唯一能看到的只有头顶的天空。客户对此感到很是震惊和感动,这种与众不同的诠释方式让他心服口服。但他还是想要一座垂直的眺望塔。因此,西扎设计了米拉多尔塔(Mirador)。随后,西扎将他的眺望塔与承孝相的玄庵和暝庭设计在同一条轴线上,这条轴线的中心切线确立了植物园的核心,具体来说:榅桲树园(Quince Garden)首先打造了植物园的中心部分,随后思潭确认了其核心位置,最后游园(Yuwon)巩固了其地位。这些建筑一个接着一个,

and the many interventions of the client himself. Approaching a decade since the first designed building was set on its ridges, Sayuwon remains an open, ongoing project, a magnificent park without a masterplan. The only possible master planner is the client whose bold moves often defy logic and calculation. What is Sayuwon? Is it a commercial venture or a personal dream? What will be its extent, in terms of both experience and physical boundary? Will the buildings have use or are they part of that rare species of useless architecture? What is clear is that Sayuwon is the epitome of humanized nature: a beautiful puzzle, maddening yet irresistible to all the different personalities involved. Yet even without a masterplan, it sustains an intelligent and sensuous organization created by a karmic interplay of its many agents.

In Sayuwon, Seung's most important capacities are both clear and manifoldly subtle. He is obviously the architect of many of its buildings, but his most important role is that of a geomancer: the client's consultant and interpreter of what, where, and how to build. Hyeonam, Sayuwon's first architectural project, was a request for a small house without any specified location. Seung found its site and created a small space that would amplify the vast mountain range into a panoramic landscape. Sadam, the next project for Seung, was a space that would simultaneously function as a café and outdoor theater. Seung responded by using an existing pond as a divider of audience and stage, all the while setting the building to mark the visitor's entry into the heart of the arboretum. The client then asked Seung to design a permanent lookout tower to replace a temporary structure at the highest point of Sayuwon. Instead of a vertical structure that the client had expected Seung proposed its antithesis, what is now called Myeongjeong – a square sunken space from which only the sky is visible. The client – simultaneously surprised, convinced, and moved by this different interpretation – still wanted a vertical tower, leading to what would become Siza's Mirador. Siza, in turn, aligned his tower on an axis with Seung's Hyeonam and Myeongjeong. It is an axis whose central tangent confirms the heart of the arboretum: the center that the Quince Garden first established, then Sadam confirmed, and finally Yuwon strengthened. It was a sequence that has set up the basic spatial structure for Sayuwon. Now from Hyeonam, Siza's tower can be seen at the left end of its panoramic view. Seung and

Welcomm City

构建出了思惟园的基本空间结构。现在,从玄庵向外眺望,左侧可以看到西扎的米拉多尔塔。承孝相和西扎对"眺望"这个词的解释虽然不同,但却也有相互交织的部分,他们二人创造了一种令人始料未及的辩证性思维。思惟园正是基于这样的理念逐步发展起来的,而承孝相最重要的能力也得到了体现,即在植物园的所有时间节奏中等待。他热切而安静地等待着。在日常生活中,他等待着做出决定,等待着各种不确定性和冲突的自我清除。承孝相的等待是几乎所有职业实践都需要的一种美德。他的等待不只是对于日常琐事的耐心,更是延伸到了人类对"等待"认知的最大范围。正是等待让他意识到人类的局限性,也让他意识到自己作品的局限性。从这个意义上来说,他的建筑作品也在等待着。

  正如我们在文章开头的引言中所提及的那样,承孝相相信人类的局限性及其奇思妙想必然灭亡的结局。这是面对不可知但确定的存在时的一种感觉。他把工作——不仅是建筑工作,还有他的社会活动——作为衡量时间的方式。当然,建筑比政治生活的节奏要慢,但他明白,二者必然是交织在一起的。尽管他的建筑属于现实社会的一部分,但我一直认为他最好的作品并不存在于教会空间中,而是存在于供人独处的空间中;但这里"独处"不是字面意义上的"独处",不是指空间中只有一个人,而是在同一空间中的人分享他们自己的孤独。他最好的作品中必定包含了能够证实人类孤独的时刻:惠康城(Welcomm City)的阶梯、惠化中心(Hyehwa)空荡荡的庭院,和某墓地的平台。每座建筑都有自己的时间节奏,都有自己独一无二的故事,但无一例外的是,这些建筑都有着"孤独"的时刻。有这样一个基本的自相矛盾的观点,那就是:我们共同的孤独证实了我们对群体的渴望。举个例子,在暝庭的庭院中,尽管池塘和各不相同的四面墙十分惹人注目,但这里的空间并非由观众和表演者、领导者和追随者所划分。在这里居住的人肯定会受到启发去读诗、奏乐、跳舞,也许还会去祈祷。人终有一死,有人在不断追求群体,而受这样的约束,我们会像知音一样在此欣赏他们的表演。尽管暝庭的空间组织似乎遵循了基督教堂的特点,包括十字形教堂的耳堂、中殿、过道和后殿,但它在宗教制度化

Siza, with their divergent but intertwined interpretation of the meaning of "looking out", has created an unforeseen dialectic. In this manner with which Sayuwon has been nurtured, Seung's most important capacity – that of waiting in all of its temporal rhythms – is illuminated. He waits both intently and with soft repose. Seung the person waits, in the everyday sense, for decisions to be made, for uncertainties and conflicts to clear itself. Seung's waiting, a virtue that is required in almost all professional practice, extends beyond the practical matters of everyday patience to the furthest extent of human cognition. It is waiting that encompasses his sense of the finitude of not only man but also of the things he produces. In this sense, his built work also waits.

As we have encountered in the quote that opened this essay, Seung believes in the profound finiteness of humanity and the mortality of its artifice. It is a sense of finitude in the face of the unknowable but certain existence. He approaches work, not only that of building but also his social activities, as a measure of time. The rhythms of building are of course longer than the vita activa of political life. But he understands that they are necessarily intertwined. Though his buildings operate as part of the practical social world, I have argued that his best work centers not on congregational spaces but on spaces meant for solitude; solitude not in the literal sense that there should be only one person in the space but in the sense that those who occupy the space share their existential loneliness. His best work inevitably contains moments that confirm the solitude of man: the in-between steps of Welcomm City, the empty courtyard of Hyehwa Center, the platform of a cemetery. Each has its own temporal rhythm, its own unique story, but they all contain moments of solitude. As a fundamental paradox, our common loneliness confirms our longing for community. For example, I believe that the courtyard in Myeongjeong, despite the temptations of the pond and the different conditions of each wall, is not a space to be divided by audience and performer, leader and follower. Occupants will certainly be inspired to read poetry, to play music, to dance, and perhaps to pray. But it is a space to enjoy these performances as comrades, bound by our mortality and search for community. And though Myeongjeong's spatial organization seemingly follows the distinctions of the Christian church – between transept, nave, aisle, and apse – it assumes a space before the institutionalization of religion, before the divisions of clergy and laity.

大田大学惠化文化中心
Hyehwa Culture Center of Daejeon University

某墓地
Graveyard

之前，在神职人员和俗人的划分之前，就已然自成空间。

思惟园的展馆建筑作为建筑作品，与绝大多数建筑项目不同。其他建筑项目是为了特定的功能而建造的，而思惟园建筑的构思和设计都是优先于特定功能的。

考虑到植物园几乎没人居住，承孝相深入研究了建筑与大气、阳光和风等季节性节奏之间的关系。尽管思惟园的建筑似乎将他对人和物有限性的看法具体化了，但却依然能在壮丽景色的诱惑和司空见惯的场景之中运作。因此，与大多数建筑日常存在的理由不同，这些建筑还存在着种种可能：让人狂妄自大；供人休闲娱乐，以弥补因为孤独而不可避免带来的焦虑。思惟园于2021年春天对外开放，如何使用这些空间现在成了一个管理和商业问题。但很容易理解的一点是，我们每天的生活一定会与社会的日常现实情况交织在一起，也就一定会产生孤独之感。

在未确定建筑的功能之前，先确定建筑的设想，这样的建筑设计顺序十分罕见。正因如此，人们很容易觉得思惟园并非是给人居住的。让我们把植物园的建筑想象成废墟。想象一下，在未来的某个时间点，也许是在地球熬过了"人类世界"的苦难之后，一群探险者冒险来到了军威郡的山中。这些开垦出来的空地将会变回原始森林。他们沿着狭窄的山谷前行，会发现曾经是卧寺的破损耐候钢。一路"披铁斩棘"后，他们将发现一个混凝土盆地，并猜测这里曾是一个人工池塘。到达其中一座山顶时，他们会看到暝庭倒塌的混凝土墙，并对其当时的用途感到困惑。他们会发现，在过去，暝庭的中间有一个空旷的广场，一个庭院，但他们不会理解其周围空间的作用。那是一口井还是用来祭天的地方？我们可以想象这些探险者会摇摆不定，摸不到头脑，暝庭是实用空间还是祭祀空间？如果他们善于解读土地，但不知道这是建筑师的匠心之作，他们可能会感觉到这些建筑的排放是很考究的。但未来的冒险者可能不知道的一件事是，这些作品即使是在刚刚建成时，看起来也非常像废墟。

The pavilions of Sayuwon, as architectural work, are exceptional in that unlike almost all architectural projects generated by the need for use, they were conceived and designed before their programs of occupation were specified.
With minimal requirements of human occupation, Seung delved into architecture's relation with the seasonal rhythms of atmosphere, sunlight, and wind. While they seemingly crystalize Seung's sense of the finitude of men and things, they operate with the temptations of spectacle and commonplace. And that is why, unlike the everyday raison d'etre of most architecture, these pavilions contain the possibility of hubris; of inviting comfort and entertainment to compensate for the anxieties that inevitably arise from loneliness. With the public opening of Sayuwon set for Spring 2021, the here-and-now question of how these spaces will be occupied has recently become a matter of management and business. The simple point here is that solitude is part of our necessary entanglement with the everyday realities of the social world.
With this rare sequence where the imagination of the building comes before its projected occupation, it is alluring to think of Sayuwon beyond its occupation by humans. Let us imagine the pavilions in Sayuwon as ruins. Imagine at some future point in time, perhaps in an era past the earth's sufferings of the Anthropocene, a group of explorers venturing into hills of Gunwi. The clearings that were created will have returned to a primordial forest. Walking along the narrow valley, they will discover the mangled corten steel of what was once Wasa. Clearing the steel and bushes, they will uncover the concrete basin and guess that this spot was a man-made pond. Reaching one of the mountain tops, they will see the crumbled concrete walls of Myeongjeong and puzzle over what it was used for. They will know that there was an empty square space, a courtyard, in the middle but will not understand what its surrounding spaces were for. Was it a well or a place to worship the sky? We can imagine these explorers oscillate between reading Myeongjeong as a practical space and a ritual space. If they are adept at reading the land, without knowing that it was the meticulous work of an architect, they may sense how carefully these structures were placed. One thing that these future adventurers may not know is that these works, even as new buildings, looked very much like ruins.

感受实用主义

Perceptive U

重新思考公共设施建筑
Rethinking the Architec

回收设施、资源中心和净水厂等公共设施有着不容置疑的实用价值,但尽管如此,绝大多数人仍然没有注意到这些公共设施。最近,一些设计师想要洗刷这些公共设施身上的"污名",决心让它们不再"隐身"。

设计师们不再接受这些设施的视觉中立性,也不再仅仅注重这些设施的效率。相反,出现了一种新的趋势。这一趋势致力于帮助我们揭秘"地下世界"——这里的"地下世界"既有字面意义,又有比喻意义——我们可以看到一些基础设施,正是这些不可或缺

Despite their unquestionable utilitarian value, public amenities such as recycling facilities, resources centers, and water purification plants remain unnoticed to the vast majority of the population. Recently, this invisibility has been challenged by designers interested in overturning the stigma associated with this kind of projects. Rather than accepting their visual neutrality and focusing only on efficiency, a new tendency is emerging, committed to unveil the underground world – literally and figuratively – of the infrastructure necessary to support the most banal acts and activities of our everyday life: lighting a lamp at home or office, taking the metro, or

哥本丘_CopenHill / BIG
坪山阳台_Pingshan Terrace / NODE Architecture & Urbanism
索勒德加德净水厂_Solrødgård Water Treatment Plant / Henning Larsen
穆滕茨净水厂_Muttenz Water Purification Plant / Oppenheim Architecture
感受实用主义：重新思考公共设施建筑_Perceptive Utilitarianism: Rethinking the Architecture of Public Amenities / Nelson Mota

tilitarianism
ure of Public Amenities

的基础设施帮助我们进行日常生活中最平常的活动：点亮家里或是办公室的灯、乘地铁，或者只是喝杯水。

这就是本节所介绍的公共设施的案例。对于建筑来说，特别是基础设施建筑，通常会具有实用主义的特征，但本节介绍的公共设施揭示了建筑如何超越这一特征。通过精挑细选的材料和巧妙的空间配置，这些建筑融入到了公共生活中，给人们带来教育、娱乐，乃至知觉方面的体验。这些建筑能给最多的人带来最大的快乐，就是"感受实用主义"的完美诠释。

drinking a glass of water.
This is the case of the projects featured in this section. They illustrate how architecture can be used to go beyond the utilitarian character usually associated to infrastructural buildings. Investing in carefully selected materials and clever spatial configurations these buildings become part and parcel of public life, providing educational, recreational, and sensorial experiences. For their capacity to bring the greatest pleasure for the greatest number of people these buildings are outstanding examples of "Perceptive Utilitarianism".

感受实用主义

重新思考公共设施建筑

Perceptive Utilitarianism
Rethinking the Architecture of Public Amenities

Nelson Mota

人们普遍认为，英国社会改革家杰里米·边沁（1748—1832年）是实用主义伦理学理论的创始人。对于边沁来说，任何事件或行为只有在能给最多的人带来最大快乐的情况下，才能定义为道德上的善。在边沁的道德理论中，我们做出任何行为的目标应该是：带来更多的快乐，带来更少的痛苦。边沁的实用主义倾向于享乐主义。他认为："'善'就是最大地增加了幸福的总量；'恶'只是增加了痛苦或者减少了快乐。"[1]

"Utilitarianism（实用主义）"这个词的词根是"utility（实用）"。根据韦氏词典，"实用"的定义是适合某种目的的东西；对某种目的有价值的东西；有用的东西；为某种用途而设计出来的东西。但奇怪的是，在商业术语中，"utility（公共设施）"的定义是拥有和/或运营一些设施的大公司，这些设施负责发电，并向公众输送或分配电力、天然气和水资源。[2]

考虑到"utility"是以上两个定义的结合，我们可以说一座"实用主义"的建筑——比如，能源设施或净水厂——应该是设计并建造出来给我们带来快乐而不是痛苦的。然而，对这些实用建筑开发项目进行投资的重要性经常遭到忽视。当我们想到这类基础设施建筑时，视觉、嗅觉或触觉的快乐绝对不会首先出现在我们脑海中。人们将大部分这类建筑视为错综复杂的系统，需要专业的工程知识，但也就仅此而已。让这类建筑避开公众的监督已经成了一种趋势，就像被锁在秘密地牢中的隐形怪物。

The English social-reformer Jeremy Bentham (1748–1832) is commonly credited as the founder of Utilitarianism as an ethical theory. For Bentham events or actions could only be judged as morally good if they result in the greatest pleasure for the greatest number of people. In Bentham's moral theory, our actions should be guided by the goal to always bring about more pleasure and less pain. Bentham's Utilitarianism is committed to hedonism. For him "goodness is just an increase in pleasure, and evil or unhappiness is just an increase in pain or decrease in pleasure".[1]

The root of Utilitarianism is the word "utility". According to the Merriam Webster dictionary, "utility" is defined as something fit for some purpose or worth to some end; something useful or designed for use. Curiously enough, in business jargon, "utility" stands for a large firm that owns and/or operates facilities used for generation and transmission or distribution of electricity, gas, or water to general public.[2]

Considering the combination of these two definitions, one could say that a "utilitarian" building – such as an energy facility or a water purification plant – should be an example of something designed and built to give us pleasure, not pain. However, the importance of investing in the architectural development of utilitarian buildings is often overlooked. Pleasure – visual, olfactive, or haptic, for example – is definitely not the first thing that comes to our mind when we think about this type of infrastructural buildings. They are mostly seen as complex systems that require expertise in engineering, but not much more than that. There is a tendency to keep this type of buildings hidden from public scrutiny, as invisible monsters locked in a secret dungeon.

Perceptive Sustainability
Buildings that accommodate infrastructural facilities are seldom objects of an integrated design

感受的可持续性

生态系统塑造了我们的建筑环境和自然环境,容纳基础设施的建筑可以将建筑的技术性能与更广泛的生态系统相结合,但这些建筑却很少成为综合设计方法的对象。然而已有明确的迹象表明,这一趋势正在发生改变。由BIG建筑师事务所设计的垃圾焚烧发电厂——哥本丘(78页)(坐落在哥本哈根),就是一个很好的例子。

耗资5.3亿欧元的阿迈厄资源中心(ARC)刚刚建成,哥本丘发电厂是其中的一个组成部分。阿迈厄资源中心包括一座可以将垃圾转化为电力的能源设施、一个滑雪坡、一面攀岩墙和一条上坡跑道。所有设施的顶部都安装了一个"智能"烟囱,每当释放的二氧化碳达到一吨之时,烟囱就会向天空中释放出环状的烟雾。阿迈厄资源中心为哥本哈根市民提供了机会,让他们在丹麦这样平坦的地区也能够拥有探索山地的体验,此外,打造这座建筑还像是一种策略,提高人们对能源的可持续使用的意识。

哥本丘被誉为近年来最具智慧的设计之一,作为一种三赢局面推广,完美解决了组成可持续性的三大元素:环境、经济和社会。哥本丘接收并处理来自大哥本哈根地区城市和企业的废物,为15万户家庭提供可回收材料、电力和集中供暖,为55万居民提供低碳电力。除了环保方面的表现,阿迈厄资源中心也被证明是一项很好的投资,因为它减少了哥本哈根对天然气和石油等不可再生能源的依赖性,并帮助哥本哈根在不远的未来成为世界上第一个碳中和之都。最后,哥本丘也为哥本哈根和周边城市的市民提供了新的娱乐选择。BIG建筑师事务所的创始人比亚克·因格尔斯解释道:"哥本丘是感受的可持续性的完美诠释——一座可持续发展的城市不仅可以改善环境,也让人们的生活变得更愉快。"[3]

approach, combining their technical performance with a wider ecosystem that shapes our built and natural environment. There are clear signs, however, this tendency is changing. CopenHill (p.78), a waste-to-energy plant designed by Bjarke Ingels Group, recently built in Copenhagen, is a case in point.

CopenHill is one of the components of the €530M new Amager Resource Center (ARC). ARC comprises an energy facility, converting refuse into electricity, a ski slope, a climbing wall, and an uphill running track, all topped with a "smart" chimney that releases smoke rings into the sky whenever one ton of fossil CO_2 is released. Next to providing opportunities for the citizens of Copenhagen to explore a mountain-like experience in a flat country such as Denmark, ARC performs also as a device to raise awareness on the sustainable use of energy.

The building gained a reputation for being one of the cleverest designs of recent years. It is indeed publicized as a win-win-win situation, addressing the three components of sustainability: environmental, economic, and social. This facility receives and processes waste from cities and businesses operating in the Greater Copenhagen area, and provides recycled materials, electricity and district heating to 150,000 households and low-carbon electricity for 550,000 people. Next to its environmental performance, ARC has also proved to be a good investment, reducing the city's dependence on non-renewal energy sources as gas and oil, and contribute to make Copenhagen the world's first carbon neutral capital in the near future. Finally, it also constitutes a new addition to the recreational options offered to the citizens of the Danish capital and surrounding cities. As Bjarke Ingels – the founder of BIG – explains, "CopenHill is a crystal clear example of perceptive sustainability – that a sustainable city is not only better for the environment – it is also more enjoyable for the lives of its citizens."[3]

净水设施

除了哥本丘发电厂之外,本节还将介绍三个净水厂项目,这三个项目也是为社会做出贡献的突出建筑案例。公共设施建筑在我们的日常生活中扮演着至关重要的角色,在探讨建筑设计怎样才能帮助改善公共设施建筑的环境质量时,这三个项目的贡献堪称典范。

获得安全饮用水可以说是人类生活最基本的需求。但尽管安全的饮用水至关重要,却并不是人人都能获取到的。事实上,联合国的数据表明,世界上有10%的人口用不上基本的饮用水装置。这就是为什么在联合国的17个可持续发展目标中,"确保人人享有可持续管理的供水和公共卫生"被列为其中之一。[4]

在发达国家,国家和地方政府的首要目标一直是提供安全的饮用水。在大多数中高收入国家,综合水资源管理在近年来已经演变成一个复杂而完善的系统。然而,在大多数中低收入国家,这种系统很不稳定,或者根本不存在。此外,随着气候变化愈演愈烈,水资源也呈现短缺趋势,水资源变得更加珍贵,更加难以获取。如果说20世纪的政治主要受石油地缘政治的影响,那么许多专家认为,水资源将成为21世纪的新型"石油"。换句话说,"水政治"将成为影响世界和平的关键因素。[5]

许多人都认为,获取安全的饮用水是理所应当的,只要打开厨房或浴室里的水龙头,水就流出来了。然而,绝大多数人都没有注意到的是,为了让我们在家里安心地喝一杯水,需要多么复杂的机制。因此,现在是时候提高人们对合理管理水资源的重要性的认识了。本节介绍的三个净水厂项目都是有意揭秘"地下世界"的典型案例——这里的"地下世界"既包含字面意义,也具有比喻意义——展现了净化我们家中饮用水所必需的基础设施。净水厂通常都摆脱不掉"实用主义"的标签,但下文要介绍的三个净水厂项目则展示了这类建筑是如何撕掉这一标签的。

Water Purification Facilities

Next to CopenHill, this section includes three projects for water purification plants, which provide other striking examples of socially relevant design. They are exemplary contributions to discuss how architecture can contribute to raise the environmental quality of public utility buildings that are vital for our everyday life.

Access to safe drinking water is, arguably, the most fundamental requirement for human life. However, despite its vital importance, safe, drinkable water is not available to everybody. In fact, according to the United Nations, ten percent of the world population do not have access to a basic drinking water service. This is one of the reasons why "ensuring availability and sustainable management of water and sanitation for all" has been placed as one of the seventeen sustainable development goals.[4]

In the developed world, the provision of safe drinking water has been a central priority for national and local governments. The management of integrated water resources has evolved over the years to become a complex, well-articulated system in most of the middle- and high-income countries. However, it is still absent or precarious in most of the low- and middle-income countries. Furthermore, with climate change, the scarcity of water tends to increase, making it even more precious and inaccessible. If the 20th century politics was chiefly influenced by the geopolitics of oil, many experts argue that water will be the new oil in the 21st century. In other words, "hydro-politics" will be the crucial factor that will influence world peace.[5]

Most of us take for granted the access to safe, drinkable water, coming out of a tap in our kitchen or bathroom. However, the complex mechanism necessary to allow us to confidently drink a glass of water at home, remains unnoticed to the vast majority of the population. It is high time, thus, to raise awareness on the vital importance of managing water resources wisely. The three projects featured in this section are exemplary cases of a deliberate attempt to unveil the underground world – literally and figuratively – of the infrastructure necessary to purify the water that we drink at home. These three water purification plants illustrate how architecture can be used to go beyond the utilitarian character usually associated to these projects.

必不可少的基础设施

南布水质净化站是当地另一项基础设施,远离市区,位于深圳市中心以东40km的坪山河岸边。成片的技术区安置在地下,地上的一侧则是中规中矩的办公大楼。在设计过程中,客户希望坪山河滨水区的开发更加一体化,并将这一项目委托给由刘珩主持的南沙原创建筑设计工作室(NODE Architecture & Urbanism),南沙工作室的办公室位于珠江三角洲地区。

南沙工作室设计的项目(第94页)探索了现有建筑的空间配置及其承载能力所带来的可能性,并为其增加了一个新的层次。该项目的主要目标之一是将水质净化站整合到沿河边分布的公共空间网络中。除了改善水质之外,这座"实用性的"建筑还能够改善附近的景观,创造新的公共空间,凭借其具有吸引力的特征融入到城市生活中。

坪山阳台(Pingshan Terrace)也是南沙工作室的项目,位于水质净化站现有办公大楼的顶部,其本质上是一个凹凸不平的屋顶,人们可以在上面行走。设计师引入了一套复杂的入口系统,用不同的形状和尺寸来连接三个楼层:两段巨型楼梯将现有办公大楼的屋顶和一层连接起来,使之与滨水区的景色遥相呼应。从办公大楼的屋顶到新建的起伏屋顶,南沙工作室创造了两个截然不同的连接方式:环绕水质净化站的烟囱所建造的环状楼梯,以及多功能的廊台,象征着通往天堂的阶梯。

南沙工作室对项目的材料和几何外形的选择似乎显得很刻意,与原有平平无奇的办公楼形成了对比。南沙工作室巧妙地整合了这些新的建筑元素,让南布水质净化站不再"隐身",为社交、公共演出或任何其他用作享乐用途的体验创造了新的可能性。

正如南沙工作室的坪山阳台为珠江三角洲设计了该地区迫切需要的公共空间,来缓解无情的城市化步伐,

Vital Infrastructure

The Nanbu Water Purification Station, located on the banks of the Pingshan River, 40kms to the east of Shenzhen's city center, was developed as yet another infrastructural construction, disconnected from the public. A series of technical areas were buried under the surface, and topped with a banal office block on one of its sides. At a certain moment of the process, the client required a more integrated development of the Pingshan riverfront and commissioned the project to NODE Architecture & Urbanism, led by Doreen Heng Liu, with offices in the Pearl River Delta region.

The project developed by NODE (p.94) explored the possibilities offered by the spatial configuration and load bearing capacity of the existing construction to add a new layer to it. One of the key goals of the project was to integrate the water purification plant in a network of public spaces to be created along the riverside. Next to improving the quality of the water, this utilitarian building would also contribute to improve the riverfront landscape, creating a new public space, with appealing features to make it part of urban life.

NODE's project, Pingshan Terrace, is essentially an undulating walkable roof, standing on top of the plant's existing office block. The designers introduced a sophisticated system of accesses, with different shapes and sizes to articulate three levels: two monumental stairs link the roof of the existing office block with the ground floor, establishing the continuity to the riverfront area. From the roof of the office block to the new undulating roof, NODE created two distinct connections: a circular stair embracing the plant's chimney, and a multi-functional tribune, that doubles as a sort of stairway to heaven.

The materials and the geometry of NODE's project seem to be deliberately chosen to contrast with the banality of the pre-existing office building. The combination of these new elements, cleverly integrated by NODE, removes the Nanbu Water Purification Station from its anonymity, and creates new possibilities for social encounters, public performances or any other hedonistic experiences.

While NODE's Pingshan Terrace creates a much-needed public space to mitigate the relentless speed of urbanization of the Pearl River Delta region, the Muttenz Water Purification Plant (p.124), designed

奥本海姆建筑设计工作室的欧洲分公司设计了穆滕茨净水厂（124页）。穆滕茨是瑞士的一个自治市，紧挨着莱茵河，位于巴塞尔以东几公里。穆滕茨净水厂顺利地整合了穆滕茨森林区域周边的大型基础设施。建筑体量经过精心设计，去除掉了与此类基础设施相关的所有熟悉特征。事实上，与传统的实用主义建筑不同，穆滕茨净水厂更像是一块岩石，其边缘形状像是数千年来不断受到自然侵蚀而形成的。净水厂的外立面和屋顶使用了喷成红色的混凝土，让人形成了这种错觉。

在设计图和剖面图中，设计师改变了由功能布局定义的传统矩形形状，进行精心设计，以适应复杂的净化系统。因此，整座建筑像是被赋予了生命，人们不经意间就能瞥见其内部环境。这个"捉迷藏"策略的高明之处是一个令人印象深刻的不规则入口，专为游客设计，看起来像设计师口中所说的壁龛。在这里，从建筑外部进入到内部的仪式感设计，让游客们屏息凝神，为紧张刺激的体验做好准备。质感、光线、声音和气味共同将游客的体验推向预期的高潮。

穆滕茨净水厂的体积并不确定，而索勒德加德净水厂（110页）同样具有这一特点。索勒德加德净水厂由丹麦的亨宁·拉森建筑师事务所设计，位于哥本哈根北部的希勒勒市，当地人口只有33 000多人。尽管索勒德加德净水厂占地12 800m²，但却几乎没有出现在自然景观中。换一种说法，索勒德加德净水厂与自然景观融为了一体。正如其设计师所说："这个位于希勒勒市的开放型净水厂有一个绿色屋顶，是野餐的绝佳去处，当地居民可以亲眼看到自己生活中所使用的资源。"

和前文提及的两个项目一样，索勒德加德净水厂积极促进公众参与到公共设施的互动体验之中，提高人们

by the European branch of Oppenheim Architecture + Design (OAD), integrates smoothly a sizeable infrastructural building in the periphery of a forested area in Muttenz, a Swiss municipality bordering the river Rhine, a few kilometers to the east of Basel. The volume of the building was carefully shaped to remove any familiar features associated with an infrastructure of this kind. Indeed, rather than a conventional utilitarian building, the Water Purification Plant of Muttenz resembles more a rock formation with edges shaped by millennia of natural erosion, an illusion stimulated by the use of red sprayed concrete in the building's outside facades and roof.

Both in plan and in section, the designers distorted and manipulated the conventional rectangular shape defined by the functional layout necessary to accommodate the sophisticated purifying system. In doing so, the building comes alive, allowing unexpected glimpses of its interior. The climax of this hide-and-seek strategy is the dramatic irregular openings that define the entrance to the area dedicated to visitors, an alcove-like room, as the designers call it. Here, the ritual of moving from the outside to the interior of the building elevates the senses, preparing the visitor for an intense experience where textures, light, sounds and scents join forces to create an expected experience of the space.

The ambiguous volumetric nature of the Muttenz Water Purification Plant, can also be found in Solrødgård Water Treatment Plant (p.110), designed by the Danish office Henning Larsen for Hillerød, a town with a little more than 33,000 inhabitants, located north of Copenhagen. Despite its 12,800m² of area, the plant almost disappears in the landscape. Or else, it becomes part of the landscape. As the designers put it "this open, green-roofed water treatment plant in Hillerød is fit for a picnic, putting the local community face to face with their use of resources".

Following the same agenda of the two projects described above, the Solrødgård Water Treatment Plant actively seeks to engage the community in an interactive experience of a public utility, raising awareness on climate change and adequate use of resources. The inner workings of the purification process accommodated in the building are only revealed in a central pathway that cuts the building in two. Through the glazed facades on both sides of the pathway, the interior becomes exposed to the visitor or passer-by. As Marie Ørsted Larsen, Senior Landscape Architect at Henning Larsen, explains,

对于气候变化以及充分利用资源的意识。在净水厂内，仅有一条中央通道可以看见净化过程的内部运作方式，这条中央通道将整个净水厂一分为二。通过通道两侧的玻璃立面，游客或行人可以看到净水厂的内部。亨宁·拉森建筑师事务所的资深景观设计师玛丽·奥斯特·拉森认为索勒德加德净水厂的设计是具有象征意义的举动，"我们希望在自然水循环和有利于社区的建造过程之间形成一种对照"。

超越实用主义

在本节介绍的项目中，设计师们顺利地解决了特定基础设施建筑和机构的特定功能需求。这些项目为它们所在地增添了新的内涵，尽管方式不同，但都令人惊叹不已。设计师对这类基础设施项目的"实用主义"表达了敬意。他们允许这些建筑中存在一些"实用主义"的内容，同时还脱掉了这些建筑身上的"隐形衣"。在这些项目中，建筑师采用了巧妙的设计决策，从而让复杂的公共设施体系与更大范围的公共空间环境融为一体。

建筑的空间布局和建材的选择在这些基础设施项目中起着关键作用。这些建筑使用富有表现力的几何形状、独特的建材，设计出了出人意料的路线和通道，这些建筑变成了刺激人们感官的机器。正是因为这一点，人们能从中获得乐趣，也提高了人们对于气候变化和充分利用资源的意识，二者正是互补的。由于这些特性，这三个净水厂项目与比亚克·因格尔斯所推崇的感受可持续性理念产生了共鸣。此外，由于这些项目能为最多的人带来最大限度的快乐，我们也可以认为这是"感受实用主义"的典范。

this becomes a symbolic gesture "to create a contrast between the natural water cycle and the constructed process that supports our communities".

Beyond Utilitarianism

In the projects featured in this section, the designers went beyond solving the specific functional requirements for specific infrastructural buildings and mechanisms. In different but equally surprising ways, these projects add new qualities to the site where they were built. They pay tribute to the utilitarian nature of this type of infrastructural programs. Not only by allowing something useful to come about, but especially by removing the layer of invisibility usually attached to them. In these projects, the architects employed smart design decisions that promote the environmental integration of a complex public utility in a larger network of public spaces.

The spatial configuration of the buildings and their materiality play a key role in these projects. Exploring expressive geometrical configurations, unusual materials, and surprising routes and pathways, these buildings become machines to activate the senses. The pleasure gained from the exposure to these sensorial features is further complemented with their capacity to raise awareness on climate change and the adequate use of resources. For these qualities, these three projects for water purification plants resonate with Bjarke Ingels' notion of perceptive sustainability. Furthermore, for their capacity to bring the greatest pleasure for the greatest number of people we could also consider them outstanding examples of "Perceptive Utilitarianism".

---

1. Mark Dimmock and Andrew Fisher, "Utilitarianism," *Ethics for A-Level*, 1st ed. (Open Book Publishers, 2017), p.14, https://www.jstor.org/stable/j.ctt1wc7r6j.5. Definition available in http://www.businessdictionary.com/definition/utility.html
2. Lizzie Crook, "BIG Opens CopenHill Power Plant in Copenhagen with Rooftop Ski Slope," *Dezeen*, October 8, 2019, https://www.dezeen.com/2019/10/08/big-copenhill-power-plant-ski-slope-copenhagen/.
3. United Nations, "Goal 6: Ensure Availability and Sustainable Management of Water and Sanitation for All," Sustainable Development Knowledge Platform, accessed December 29, 2015, https://sustainabledevelopment.un.org/sdg6.
4. Bryan Lufkin, "Why 'Hydro-Politics' Will Shape the 21st Century," accessed July 5, 2020, https://www.bbc.com/future/article/20170615-why-hydro-politics-will-shape-the-21st-century.

哥本丘
CopenHill

BIG

集滑雪和发电为一体的标志性建筑
Skiing and power generation combine in a landmark hybrid structure

哥本丘发电厂,也叫阿迈厄·巴克,距丹麦首都哥本哈根的中心仅两公里远。哥本丘是一座标志性建筑,有一根悬挑的烟囱,烟囱旁边有一个独特的屋顶,屋顶向下倾斜,转个弯直达地面。哥本丘占地41 000m²,是一家新型的垃圾发电厂,其顶部有一条滑雪坡、一条徒步小径和一面攀岩墙,在体现感受主义可持续性概念的同时,哥本丘发电厂的目标是截止到2025年成为世界上第一座碳中和城市,这与哥本哈根的目标是一致的。2019年,哥本丘作为哥本哈根的城市娱乐中心对公众开放,其发电厂的部分在2017年就已经开始投入使用。该项目由丹麦BIG建筑师事务所设计,由哥本丘负责休闲设施的运营,而阿迈厄资源中心(ARC)负责废物管理和能源生产。

从一开始,哥本丘的定位就是:一个可能会对社会产生副作用的公共基础设施。但哥本丘取代了附近一家有50年历史的垃圾焚烧厂,哥本丘的新型垃圾焚烧设备整合了最新的垃圾处理和能源生产技术。哥本丘位于阿迈厄的滨水工业区,现在,这里的工业原料生产设施已经成为极限运动的场地,从水上滑板到卡丁车,这座新建的发电厂为该地区增加新的运动项目带来了可能性。

发电厂的机械装置根据高度进行了精准的定位和组织,设计师据此最终确定了室内空间的规模,并为9000m²的滑雪斜坡建造了倾斜的屋顶。滑雪高手们拥有长度足以媲美奥运会U形滑道的人造斜坡,斜坡表面铺着绿色的塑胶,他们可以从滑雪场顶部一直向下滑,试一试自由式或障碍式滑雪动作。而孩童和初学者可以在较低的初级坡道练习。滑雪者在乘升降机(将滑雪者送至坡顶的缆车系统)、地毯升降机(传送带)或观光电梯前往公园坡顶的时候,可以看到发电厂废物焚烧炉内24小时不间断的运行状况。

当游客们来到哥本丘的顶端时,他们将感受到"平地国度"丹麦所没有的新奇山地体验。除了滑雪以外,屋顶还设有露天酒吧、健身、攀

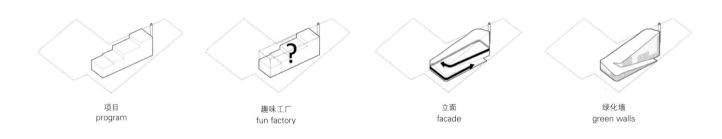

项目 program   趣味工厂 fun factory   立面 facade   绿化墙 green walls

Situated just two kilometers from the center of the Danish capital Copenhagen, the CopenHill power plant, also known as Amager Bakke, is an architectural landmark defined by a cantilevered chimney beside which a unique roof slopes down and turns to reach the ground level. The 41,000m² facility is a new breed of waste-to-energy plant topped with a ski slope, hiking trail and climbing wall, embodying the notion of perceptive sustainability while aligning with Copenhagen's goal of becoming the world's first carbon-neutral city by 2025. CopenHill opened as an urban recreation center in 2019, superimposed on the plant building which became operational in 2017. Designed by Danish practice BIG (Bjarke Ingels Group), the project is operated by Copenhill, responsible for the recreational facilities, and ARC, responsible for waste management and energy production.

CopenHill was conceived as a public infrastructure with intended social side-effects from day one. Replacing an adjacent 50-year old waste-to-energy plant, CopenHill's new waste incinerating facilities integrate the latest technologies in waste treatment and energy production. Due to its location on the industrial waterfront of Amager, where raw industrial facilities have become the site for extreme sports from wakeboarding to go-kart racing, the new power plant adds new sporting possibilities to the area.

The internal volumes of the power plant are determined by the precise positioning and organization of its machinery in height order, creating an efficient, sloping rooftop fit for a 9,000m² ski terrain. At the top, experts can glide down the artificial ski slope surfaced with green plastic and with the same length as an Olympic half-pipe, test the freestyle park or try the timed slalom course, while beginners and children practice on the lower slopes. Skiers ascend the park from the platter lift (a cable system which transports skiers uphill), and carpet lifts (conveyor belts) or glass elevator allow for a glimpse inside the 24-hour operations of a waste incinerator.

Visitors reaching the summit of CopenHill will feel the novelty

岩和全市最高的观景台。随后，游客们还可以沿着490m长、绿树成荫的徒步和跑步路线缓缓下山，沿途感受丹麦景观设计事务所（SLA）设计的茂密山林。同时，面积达10 000m²的绿化屋顶还在一定程度上解决了这个85m高的公园的微气候所带来的挑战。在吸收多余热量、去除空气微粒和调整雨水径流使其最小化的同时，也带来了生物多样性的景观。

在屋顶斜坡的下方，熔炉和涡轮机每年将44万吨废物转化为足够清洁的能源，为15万户家庭提供电力和集中供热。为达到这一目标，设计师必须对发电厂的通风井和进风口处进行特别的考量，这就为"山坡"多样化地形的创造提供了可能。因此，"山坡"作为一个人造景观，可以说是底层需求与顶层期待之间碰撞的产物。同时，阿迈厄资源中心占据了发电厂内10层的行政管理空间，包括一个600m²的环境教育中心，用于学术参观、工作坊和可持续发展会议。烟囱悬挂在主体建筑的外部，位于顶部的观景台附近。烟囱高达123m，让哥本丘成为哥本哈根最高的建筑。

与其说哥本丘是一座孤立的建筑，其实它的建筑外围护结构倒更像是一个契机，帮助改善周围的工业建筑环境，同时也成为人们旅游休闲的目的地。哥本丘的连续立面由1.2m高、3.3m宽的铝砖（看起来像巨大的砖块）堆叠而成，玻璃窗户则镶嵌在缝隙之间，让阳光能够入射到发电厂深处。同时，设计师还在西南立面设计了一些较大的洞口，为阿迈厄资源中心的行政工作区提供照明。此外，85m高的攀岩墙像是一条精心设计的垂直带，位于发电厂最长的垂直立面上。这面全世界最高的人造攀岩墙，让登山者在登山时可以一睹工厂内部的景象。在滑雪坡道的底部，设计师还设置了一个600m²的"滑雪酒吧"，随时欢迎市民和参观者的到来。

在哥本哈根体验高山滑雪
alpine skiing in Copenhagen

三条滑雪道
three pists

滑雪坡入口
slope access

在公园里散步
take a walk in the park

of a mountain in Denmark, which is essentially flat. Non-skiers can enjoy the rooftop bar, fitness training area, climbing wall or highest viewing plateau in the city before descending the 490m tree-lined hiking and running trail within a lush, mountainous terrain designed by Danish Landscape Architects SLA. Meanwhile, the 10,000m² green roof, which borders the ski slope, addresses the challenging micro-climate of an 85m-high park. It offers a biodiverse landscape while absorbing heat, removing air particulates and minimizing stormwater runoff. Beneath the slopes, furnaces and turbines convert 440,000 tons of waste annually into enough clean energy to deliver electricity and district heating for 150,000 homes. The necessities of the power plant to complete this task, from ventilation shafts to air-intakes, help create the varied topography of a mountain – a man-made landscape created in the encounter between the needs from below and the desires from above. ARC occupies ten floors of administrative space, including a 600m² environmental education center for academic tours, workshops and sustainability conferences. The chimney flue is suspended and external to the main volume, and is sited near the viewing platform at the summit. It reaches a height of 123m, making Copenhill Copenhagen's tallest building. Rather than considering Copenhill as an isolated architectural object, the building envelope is conceived as an opportunity for the local industrial context while forming a leisure destination. CopenHill's continuous facade comprises 1.2m tall and 3.3m wide aluminum bricks stacked like gigantic bricks overlapping with each other. Inbetween, glazed windows allow daylight to reach deep inside the facility, while larger openings on the southwest facade illuminate workstations on the administrative floors. Emerging as an organically-contoured vertical strip on the longest vertical facade, an 85m-high feature is designed to be the tallest artificial climbing wall in the world, offering climbers views inside the factory. At the bottom of the ski slope, a 600m² après-ski bar welcomes locals and visitors.

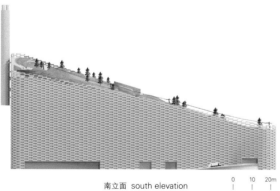

南立面 south elevation

0 10 20m

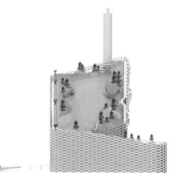

东立面 east elevation

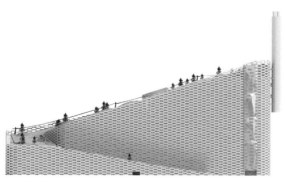

北立面 north elevation

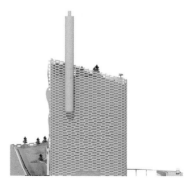

西立面 west elevation

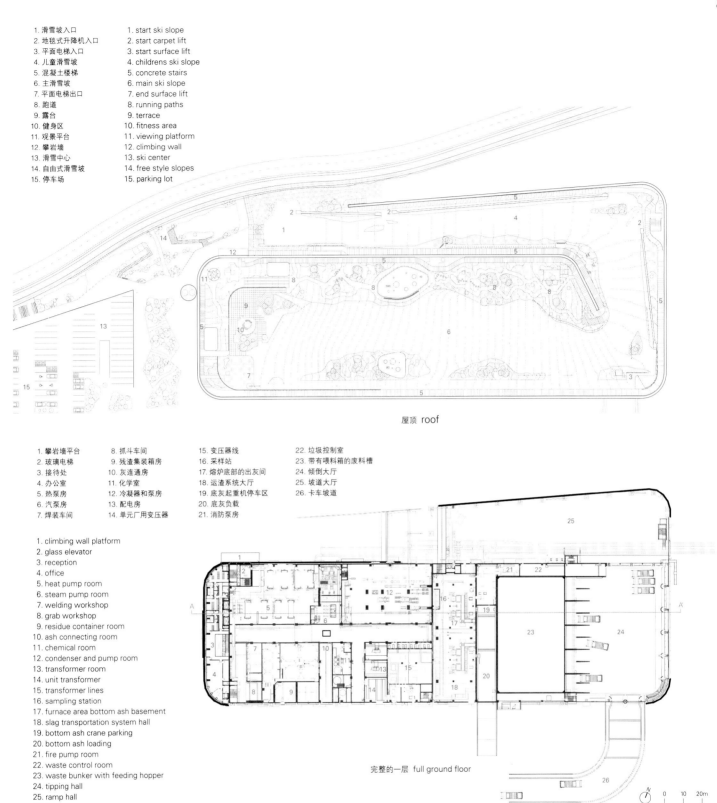

1. 滑雪坡入口	1. start ski slope
2. 地毯式升降机入口	2. start carpet lift
3. 平面电梯入口	3. start surface lift
4. 儿童滑雪坡	4. childrens ski slope
5. 混凝土楼梯	5. concrete stairs
6. 主滑雪坡	6. main ski slope
7. 平面电梯出口	7. end surface lift
8. 跑道	8. running paths
9. 露台	9. terrace
10. 健身区	10. fitness area
11. 观景平台	11. viewing platform
12. 攀岩墙	12. climbing wall
13. 滑雪中心	13. ski center
14. 自由式滑雪坡	14. free style slopes
15. 停车场	15. parking lot

屋顶 roof

1. 攀岩墙平台	8. 抓斗车间	15. 变压器线	22. 垃圾控制室
2. 玻璃电梯	9. 残渣集装箱房	16. 采样站	23. 带有喂料箱的废料槽
3. 接待处	10. 灰连通房	17. 熔炉底部的出灰间	24. 倾倒大厅
4. 办公室	11. 化学室	18. 运渣系统大厅	25. 坡道大厅
5. 热泵房	12. 冷凝器和泵房	19. 底灰起重机停车区	26. 卡车坡道
6. 汽泵房	13. 配电房	20. 底灰负载	
7. 焊装车间	14. 单元厂用变压器	21. 消防泵房	

1. climbing wall platform
2. glass elevator
3. reception
4. office
5. heat pump room
6. steam pump room
7. welding workshop
8. grab workshop
9. residue container room
10. ash connecting room
11. chemical room
12. condenser and pump room
13. transformer room
14. unit transformer
15. transformer lines
16. sampling station
17. furnace area bottom ash basement
18. slag transportation system hall
19. bottom ash crane parking
20. bottom ash loading
21. fire pump room
22. waste control room
23. waste bunker with feeding hopper
24. tipping hall
25. ramp hall
26. truck ramp

完整的一层 full ground floor

项目名称：CopenHill / 地点：Amager, Copenhagen, Denmark / 建筑师：BIG / 主管合伙人：Bjarke Ingels, David Zahle, Jakob Lange, Brian Yang / 项目主管：Jesper Boye Andersen, Claus Hermansen, Nanna Gyldholm Møller / 项目团队：Alberto Cumerlato, Aleksander Wadas, Alexander Codda, Alexander Ejsing, Alexandra Gustafsson, Alina Tamosiunaite, Armor Gutierrez, Anders Hjortnæs, Andreas Klok Pedersen, Annette Jensen, Ariel Wallner, Ask Andersen, Balaj Ilulian, Blake Smith, Borko Nikolic, Brygida Zawadzka, Buster Christensen, Chris Falla, Chris Zhongtian Yuan, Daniel Selensky, Dennis Rasmussen, Espen Vik, Finn Nørkjær, Franck Fdida, Gonzalo Castro, Gül Ertekin, George Abraham, Helen Chen, Henrick Poulsen, Henrik Rømer Kania, Horia Spirescu, Jakob Ohm Laursen, Jean Strandholt, Jelena Vucic, Jeppe Ecklon, Ji-young Yoon, Jing Xu, Joanna Jakubowska, Johanna Nenander, Kamilla Heskje, Katarzyna Siedlecka, Krzysztof Marciszewski, Laura Wätte, Liang Wang, Lise Jessen, Long Zuo, Maciej Zawadzki, Mads Enggaard Stidsen, Marcelina Kolasinska, Marcos Bano, Maren Allen, Mathias Bank, Matti Nørgaard, Michael Andersen, Narisara Ladawal Schröder, Niklas A. Rasch, Nynne Madsen, Øssur Nolsø, Pero Vukovic, Richard Howis, Ryohei Koike, Se Hyeon Kim, Simon Masson, Sunming Lee, Toni Mateu, Xing Xiong, Zoltan David Kalaszi, Tore Banke, Yehezkiel Wiliardy / 合作方：Detailed design _ SLA, Lüchinger+Meyer, MOE, Rambøll, Jesper Kongshaug and BIG Ideas; Competition _ AKT, Topotek 1, Man Made Land, Realities:United; Thanks to _ A.P Moller Fonden, Lokale - og Anlaegsfonden, Nordea Fonden, Fonden R98, Københavns Kommune, Frederiksberg Kommune, Tårnby Kommune, Dragør Kommune og Hvidovre Kommune / 客户：Amager Ressourcecenter / 功能：body culture / 面积：41,000m² / 竣工时间：2019 / 摄影师：©Rasmus Hjortshoj (courtesy of the architect) - p.82, p.84~85, p.86, p.87, p.88, p.92~93; ©Laurian Ghinitoiu (courtesy of the architect) - p.78~79, p.83; ©Soren Aagaard (courtesy of the architect) - p.91 lower; ©SLA (courtesy of the architect) - p.91 upper

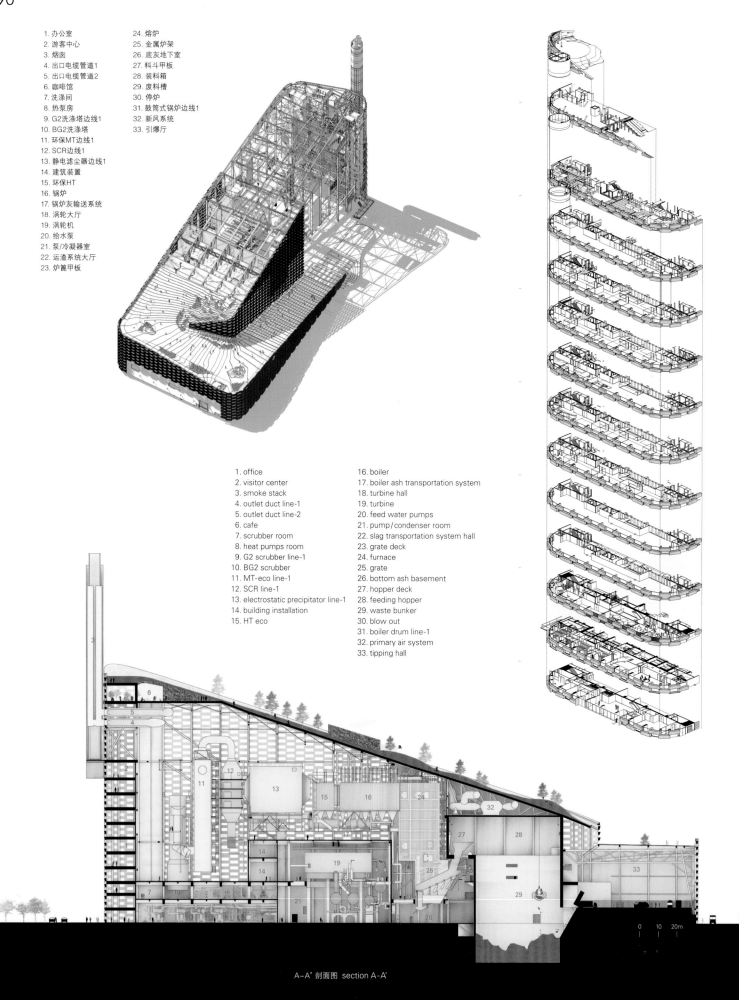

A-A' 剖面图 section A-A'

坪山阳台
Pingshan Terrace

NODE Architecture & Urbanism

漂浮在净水基础设施之上的雕塑公共空间
Sculptured public space floats above water purification infrastructure

南布水质净化站位于深圳东北部的坪山区，坐落在坪山河畔。在过去，坪山区的高速发展也给坪山河带来了污染。南沙原创建筑设计工作室（NODE）是深圳的一家建筑事务所，刘珩是事务所的负责人。南沙工作室设计了面向公众开放的坪山阳台。坪山阳台横跨并覆盖了南部水质净化站这个市政设施。

在得到扩建公共空间的指令时，南沙工作室已经建造完成了5100m²的地下水质净化站的混凝土框架。另外还要在地上建造1200m²的办公室。设计挑战是：在对称的单层建筑物上方有效地增加新的开放公共空间，并确保两者的交通流线是分离的。

坪山阳台提供了一条新的公共通道，从公园直通北面的办公建筑屋顶。在南面，梯田通向一个地上广场，广场上有四个矩形湖泊，构成水景，河流继续延伸，流入坪山河。这些湖泊的水从地下水质净化站抽取，然后排入坪山河。这条通道借助楼梯和楼梯间笔直的过渡平台，从北面的办公建筑屋顶直通南面。

在办公建筑的屋顶上，有一个钢质的阶梯平台，与湖泊和周围的景观相呼应，形成一个构件，是整座建筑的雨篷，也是屋顶公共区域，有其独特的景观造型。阶梯平台由多维折叠平面构成，形成了新的浮动式屋顶，并与二层的办公屋顶交织在一起，丰富了空间体验。该建筑面向公众开放，允许公众驻足停留欣赏，同时，其独立的步行系统与水质净化站的运行和管理是分开的。

在水质净化站已有的混凝土框架下，若是要创造新的建筑，轻钢结构则是唯一的选择。但其中一个严峻的挑战在于：多出来的不规则空间和折叠平面的屋顶，要与现有规则的混凝土结构柱网相呼应。浮动屋顶由桁架构成，桁架则由规则的钢柱网格支撑，该桁架与水质净化站现有的混凝土结构柱网相连。二层上方天花板的钢结构清晰可见，这也是新露台的底面。

落在折叠平面上的雨水流到屋顶边缘，屋顶开有一条排水沟，随后，雨水可以通过立管排入下面的市政管道网络。屋顶檐口的最低点位于屋顶四周的角落，建筑师将其设计成一个天然的排水口，将雨水直接排入建筑外的地面水景湖泊，既实用又美观。

屋顶结构由一层一层的钢架构成。主骨架（坚实的基础层）外面包裹着次骨架，二者都表现出钢架所体现出来的折叠感。顶层为塑化

木材，有接缝，便于排水。顶层的地板表面看起来像木板路，一尘不染，踩上去令人心情愉悦，同时还装有安全护栏。

二层可以作为公共剧场，这里还包含了一个封闭的镶嵌玻璃的空间，位于折叠屋顶的下方，作为活动平台的一部分。层层台阶从二层一直向上，通往折叠的空间当中去。从坪山阳台的屋顶可以俯瞰邻近的湿地公园，将坪山河及城市的全景尽收眼底。

通风井距地下建筑高15m。一个螺旋楼梯将通风井接入坪山阳台的屋顶，将新的屋顶平台和二层连接起来。这个螺旋楼梯给游客提供了另一条路线，是屋顶的又一个亮点。

坪山阳台的设计凸显出流动性和连续性，注重体验，既忠实于建筑原本的功能，又提供了新的公共空间，与当地的气候特征相融合。坪山阳台成了南沙工作室设计基础设施的新作品，它不仅是建筑空间，更是城市生活的公共空间。从精神层面和审美角度出发，可以将坪山阳台重新定义为一个以人为本的基础设施，追求我们日常生活中的"仪式感"。

The Nanbu Water Purification Station in the Pingshan District of north-east Shenzhen, is sited on the Pingshan river, which had been polluted by the area's rapid development. NODE, a Shenzhen-based architectural practice led by Doreen Heng Liu, has designed a publicly-accessible structure called Pingshan Terrace that traverses and crowns this municipal facility.

The concrete frame for the 5,100m² underground water purification plant had already been completed when NODE was asked to add public space. A further 1,200m² of offices would be above ground. The challenge was to effectively insert new public open space into this symmetric one-floor outcrop and ensure a separated circulation.

Pingshang Terrace provides a new public passage from park space to the north to the office roof. To the south, it connects to a ground-level square bordered by four water features in the form of rectangular lakes, and continues to the Pingshan

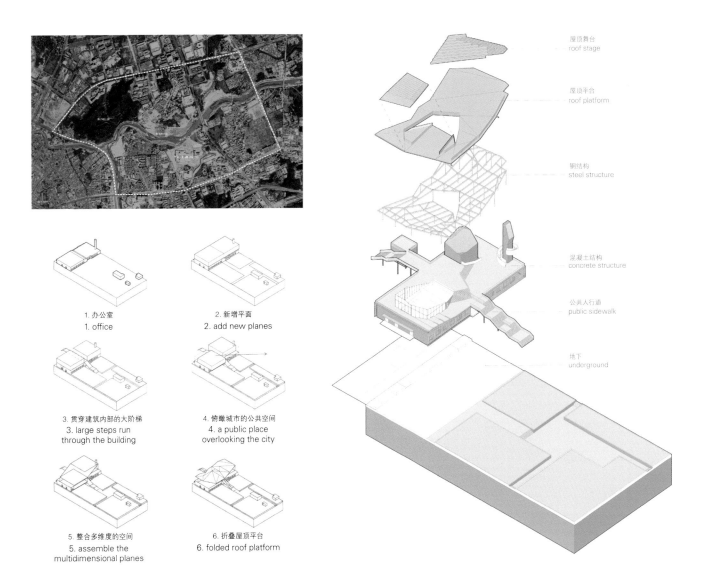

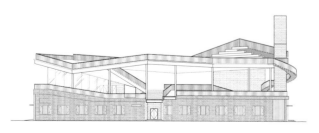

0 5 10m   南立面 south elevation

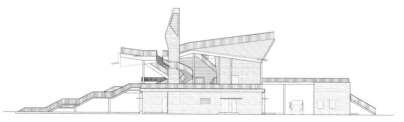

东立面 east elevation

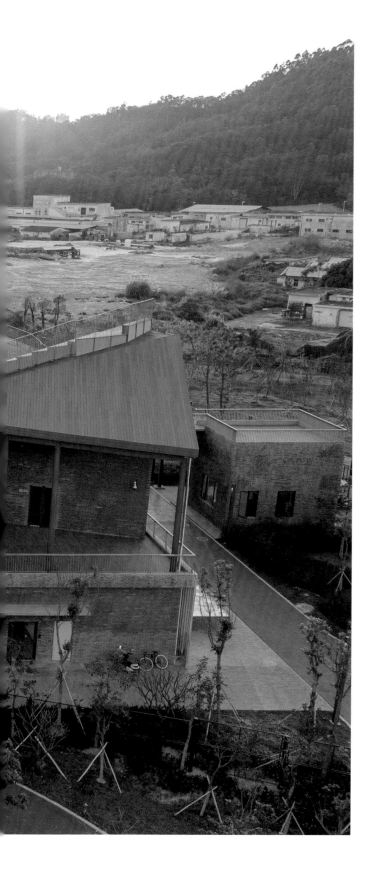

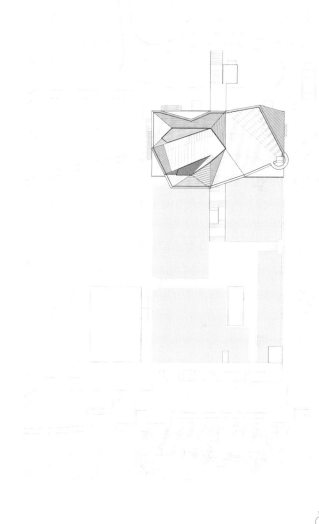

北立面 north elevation

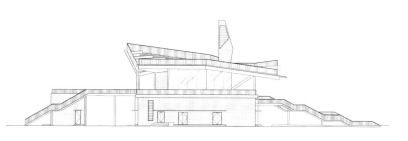

西立面 west elevation

River. The water of these lakes is drawn from the underground purification station before drainage into the Pingshan River. The route traverses the office roof via stairs directly from the north and a sequence of stairs and linear landings to the south.

Above the office roof, steel is structured to form a stepped platform which responds to the lakes and surrounding landscape, forming a structural element that is simultaneously a canopy over the whole building and a public roof area with its own landscape topography. The platform comprises multi-dimensional folded planes which form the new floating roof, and are intertwined with the office roof at level F2, enriching the spatial experience. While allowing citizens to stay and observe, the structure's independent system of pedestrian circulation is separated from the operation and management of the water purification plant.

A light steel structure was the only option for creating the new structure on the plant's concrete frame. One of the major challenges was to have the additional irregular space and the roof of folded planes correspond to the regular structural column grid of the existing concrete. The floating roof is formed by a truss supported by a regular steel column grid which connects with the structural column grid of the plant's existing concrete. The steelwork is visible in the ceiling above level F2, which is the underside of the new terrace level.

Rain falling on the folded planes flows to the roof edge where it is channeled by a gully, then drained via the rain riser into the municipal pipeline network below. The lowest point of

the roof eave, situated at a perimeter corner, is designed as a natural drainage outlet to discharge into the ground-level water feature lakes outside the building in a direct, useful and aesthetic way.

The roof structure builds up from the steel frame in layers. A main keel, or solid base layer, is overlaid with a secondary keel, both expressing the folds defined by the steel frame. The top layer is made of plastic wood, with seams to facilitate drainage. It produces a floor surface that looks like boardwalk with a clean, pleasant finish. Safety guardrails are installed.

The F2 level can serve as a public theater and includes an enclosed, glazed space under the folded roof as part of its activity platform. A large area of steps rises from F2 into the folded through an open void in it. The roof of Pingshan Terrace overlooks an adjacent wetland park and offers panoramic views of the Pingshan River and the city.

A ventilation shaft rises 15m higher from the underground building. A spiral staircase is provided to incorporate the shaft into Pingshan Terrace's roof system, connecting the new upper platform and F2. It offers another visitor route and is a high spot of the roof system.

Pingshan Terrace is a continuous, flowing experience-based interface that is simultaneously a functional building and a structural public space celebrating local climate features. It adds to NODE's recent portfolio of infrastructure that is not only engineered space but also public space for urban life. It is redefined spiritually and aesthetically as human-oriented infrastructure to pursue rituality in daily life.

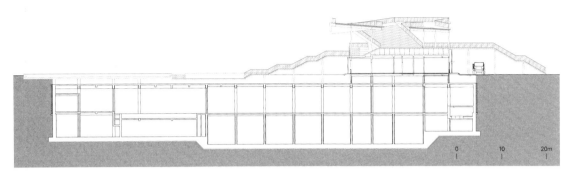

A-A' 剖面图 section A-A'

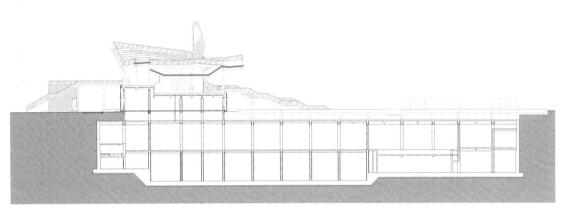

B-B' 剖面图 section B-B'

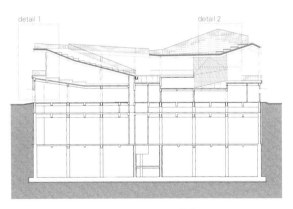

C-C' 剖面图 section C-C'

项目名称：Pingshan Terrace / 地点：Pingshan District, Shenzhen, China / 建筑师：NODE Architecture & Urbanism / 总建筑师：Doreen Heng Liu
设计团队：Doreen Heng Liu, Jiebin Huang, Xinjie He, Shihan Zhang, Chen Lian, Qingsong Lu, Xueshi Chang (intern) / 施工：Water Authority, Pingshan district, Shenzhen / EPC合同：contract :China State Construction Engineering Corporation / 施工图：CSCEC AECOM Consultants Co., Ltd
景观：NODE Architecture & Urbanism, CSCEC AECOM Consultants Co., Ltd-Shenzhen branch / 地下净化站及湿地工程设计：CSCEC AECOM Consultants Co., Ltd / 用途：water infrastructure, public space, supporting facilities / 用地面积：9,500m² / 总建筑面积：GF office area _ 1,200m²; 2F educational & exhibition & supporting area _ 280m², platform area _ 920m²; roof platform area _ 1,280m² / 结构：concrete (original), steel (new)
材料：brick, plastic wood, steel, glass / 设计时间：2017.12－2018.10 / 施工时间：2018.12－2019.7
摄影师：©Chao Zhang (courtesy of the architect)

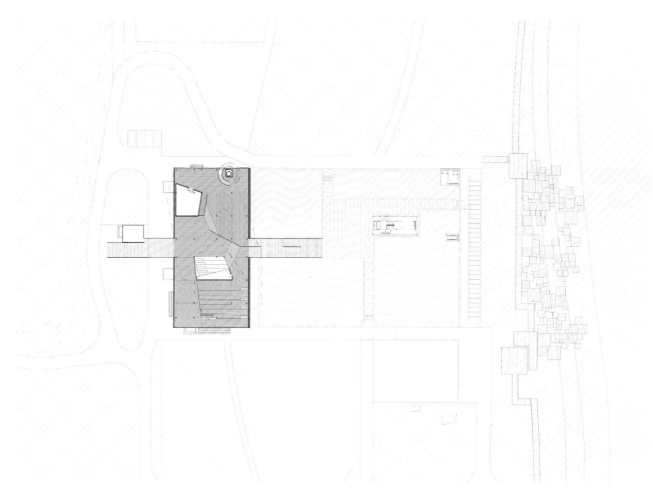

二层 first floor

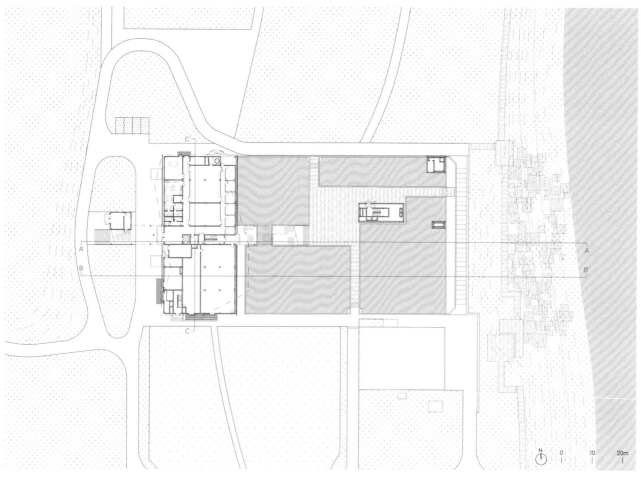

一层 ground floor

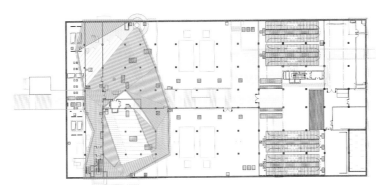

三层 second floor

索勒德加德净水厂
Solrødgård Water Treatment Plant

Henning Larsen

融入到景观中的净水厂
A water treatment plant is embedded in the landscape

索勒德加德净水厂将其工业形式和污水的气味隐藏在绿色屋顶之下,几乎隐身于丹麦的风景之中。但是,索勒德加德净水厂也引起了人们对供水相关问题的关注。每年至少有一个月,世界上近三分之二的人口要经历严重的水资源短缺。而随着气候变化加剧和全球人口逐渐拥挤,缺乏水资源人口的数量只会变得越来越多。

索勒德加德气候与环境公园坐落在丹麦的希勒勒市,旨在引起人们对于可持续资源利用这一全球性挑战的注意。索勒德加德气候与环境公园从一个占地50ha、投资10亿丹麦克朗(1.5亿美元)的总体规划演变而来,旨在通过创建对公众有吸引力的市政基础设施,就资源利用和气候意识开展社区对话。亨宁·拉森建筑事务所对索勒德加德净水厂做的外观设计和景观美化为这座开放公园做出了贡献,他们也将这一公共设施融入到一个开放的框架当中。

公园里有一个回收中心、废水处理厂和行政管理设施,旁边是步道、一座观鸟塔以及一个为当地蝙蝠准备的栖息设施。索勒德加德公园将休闲空间融入到公共设施中,创造了一个独特的空间,游客可以在这里亲身体验社区内自然资源的循环过程。资源短缺对发展中社区的影响尤为明显,而这样的设计也扩大了人们对资源短缺这一状况的批判性讨论。

索勒德加德气候与环境公园将工业设施融入到景观中,从概念上讲,就是将草地抬高,将工业设施的功能放入地下。一条便道将整体空间一分为二,看起来像是一条明沟,穿过整个景观。"明沟"中暴露在外的立面上都覆盖着耐候钢,这让建筑外部看起来就像土地被分割开的实体剖面切口。顶部的屋顶是蜿蜒的步道网络,可以纵观整个公园的景色。这些步道在栽满了多年生植物的植床上相互交错,植床上花朵的颜色随着季节的变化而变化(绿色、黄色和红色)。中央切口两边的道路网络由一根钢桥连接,由钢铁制成的楼梯从屋顶向下延伸到切口地面上。屋顶有一些钢顶的楔形天窗,游客可以透过这些天窗一瞥工厂的废水处理和过滤设施,观看净水厂是如何日均处理1.5万立方米废水的。

这一设计理念让社区与自身的资源利用联系起来,同时尽可能掩藏净水厂的存在,消除其气味。中央通道让游客了解他们社区的水循环是如何运作的,并让他们了解自然水循环和服务社区的人造水循环

有何区别。路边的一条小溪缓缓流过中央通道,流经一个狭窄的花园,展示了天然植物是如何清洁和过滤地下水的。对于经过这里的游客来说,天然植物的清洁过滤特点与净水厂的工业外观形成了另一种对照,促使人们反思公共设施的功能和环境足迹。

索勒德加德净水厂能进一步提高水处理能力,从而支持希勒勒市及其周边地区未来的进一步发展,同时还可以回收磷元素,通过废水产生生物质热能。未来的社区应该对这样一种景观模式加以探索,既可供人休闲娱乐,创造人与自然的联系,又有教育意义,唤醒人们的气候意识。

Almost invisible in the Danish landscape, the Solrødgård Water Treatment Plant hides both its industrial form, and the smell of wastewater, under a green roof. However, it draws attention to issues related to water supply. Nearly two-thirds of the world's population experiences severe water scarcity during at least one month of the year. As climate change and global crowding intensify, this figure will only become more severe.
The Solrødgård Climate and Environment Park, serving the Danish city of Hillerød, seeks to shine a spotlight on the global challenge of sustainable resource use. Developed from a 50-hectare, 1 billion-DKK (US$ 150 million) masterplan, the park aims to open a community dialogue on resource use and climate awareness by creating public appeal within municipal infrastructure. Henning Larsen contributed to the open park with their exterior design and landscaping for the Solrødgård Water Treatment Plant, embedding the public facility within an accessible earthen framework.

Here, a recycling center, wastewater treatment plant and administrative facility stand alongside walking trails, a birdwatching tower, and a roosting facility for local bats. By weaving recreational space into public utilities, the park creates a unique space where visitors can gain a natural, firsthand exposure to the cycle of natural resources within the community. The de-

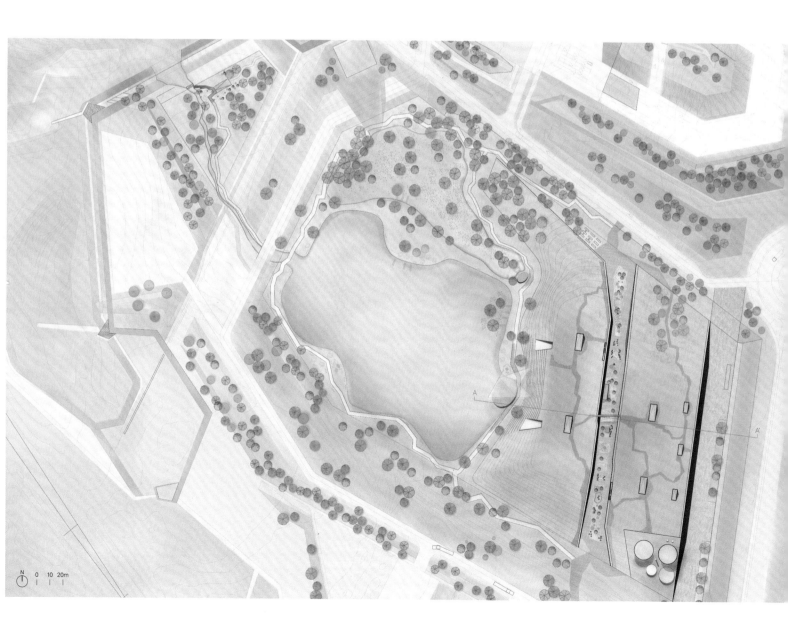

sign extends a critical conversation on resource scarcity, which disproportionately affects developing communities.
The project inserts the industrial facility into the landscape, conceptually lifting the grass and up to insert its program in the space underneath. A service road splits the space in two, running in what appears to be an open trench cut through the landscape. The facades exposed in this cut are clad in weathering steel, giving the exterior the appearance of a solid sectional cut of the earth. The roof on top is a web of meandering pedestrian paths with views across the park. They weave through perennial plant beds whose blossoms shift in color through the seasons (green, yellow, and red). The path networks either side of the central cut are connected by a single steel bridge over it, and steel staircases descend to it from the rooftop. Visitors can peer through steel-topped wedge-shaped skylights on the roof into the plant's processing wing and filtration facilities, to watch how the plant treats 15,000 cubic meters of wastewater each day.
The design concept allows the community to connect with their own use of resources, while minimizing the visual and

olfactory presence often associated with water treatment plants. The central pathway gives visitors an idea of how their community's water cycle works, and creates a contrast between the natural water cycle and the constructed process that serves the community. A small creek beside the road trickles through this central channel, passing through a narrow garden that demonstrates how natural foliage cleans and filters groundwater. For visitors passing through, this feature is another contrast to the industrial presence of the water treatment plant, prompting reflection on the function and environmental footprint of the public utility.

The Solrødgård Water Treatment Plant is capable of expanding its processing capacity to support future growth in and around Hillerød, and is capable of recycling phosphorous and producing biomass heat energy from the wastewater. Future communities will be able to explore a landscape that provides recreation, a connection to nature, and an everyday education in climate awareness.

项目名称：Solrødgård Water Treatment Plant
地点：Lyngevej 2, 3400 Hillerød, Denmark
建筑师：Henning Larsen
项目经理：Anders Park
项目团队：Annesofie Feidenhansl Milner, Daniel Baumann, Jacob Astor, Lærke Dyrholm Møldrup, Marie Ørsted Larsen, Melissa Sandoval, Mikkel Hune, Morten Hauch, Nikolaj Sandvad Ramskov, Omar Dabaan
总体规划建筑师：Gottlieb Paludan
工程师：Orbicon
承包商：Jakobsen & Blindkilde
客户：Hillerød Forsyning
建筑面积：12,800m²
设计时间：2015—2017 / 施工时间：2015—2017
摄影师：
©Jacob Due (courtesy of the architect) - p.110~111, p.112, p.116, p.117, p.119, p.120~121
©Daniel Baumann (courtesy of the architect) - p.114~115, p.118, p.122~123

立面图 elevation

A-A' 剖面图 section A-A'

穆滕沃净水厂
Muttenz Water Purification Plant
Diogenheim Architecture

一座揭秘水质净化内部运作方式的有机建筑
An organic building reveals the inner life of water purification

森林旁边,坐落着一座外形不规则的建筑,像是从地面凸起的有机岩层,岩层上还有着洞穴一样的洞口。这座建筑就是奥本海姆建筑设计公司设计的全新市政净水厂,瑞士的穆滕茨市将这一任务委托给了奥本海姆建筑设计公司,希望这一设施可以为市民提供生活饮用水。穆滕茨净水厂十分注重保护莱茵河沿岸的环境,是可持续发展的典范。净水厂位于受保护的森林和附近的工业园区之间,甚至还包括教育功能,从而在这样紧张的环境之下也能展示复杂的净水过程。

穆滕茨净水厂有一个独特且重要的功能,那就是成为穆滕茨和巴塞尔地区的一座地标建筑,该建筑设计的作用是将独特而先进的技术放入自然生态系统中,强调了净水过程的重要性。

穆滕茨净水厂位于莱茵河边一片苍翠繁茂的森林中。这种自然环境和工业环境对照鲜明的大环境通过建筑理念得以实现。

净水厂内部的运作方式是以工程学为导向的,这也决定了净水厂的外形和体积。就像一件紧身连衣裙,皮肤紧紧贴着裙子,向外界展示净水厂内充满科技感的运作方式。管道、过滤器和设备以抽象的方式呈现在净水厂的立面上。这样的设计方式呈现出了一座富有表现力的建筑。净水厂背靠大自然,就像一个"物体宝库"(自然艺术品),简化至只剩下其建材与外形。这座净水厂拥有最先进的三段净水处理技术,确保穆滕茨市的市民享用最高品质的饮用水。净水厂坐落在公共区域,随时欢迎市民来欣赏复杂的净水过程,全身心赞美饮用水。通过引导游客穿过不同的建筑区域,领略不同的净水工艺流程,也实现了理想的教学效果。净水厂还有一个开放的平台,负责收集雨水,向人们展示净水的过程。

这个像壁龛一样的房间坐落在水池上,一尘不染,向外开放,反射日光,收集从屋顶倾泻进来的雨水。根据时间和季节的不同,这个房间将会变得潮湿而寒冷,而且总是带有一丝神秘的色彩。人们对水有了前所未有的全面体验。大环境的两种对立状态在立面材料中得到了有趣的体现:轮廓柔和,材质却粗糙而坚硬。房间的四面和屋顶均采用喷射混凝土材质,使立面更具表现力。喷射混凝土采用多孔渗透的处理方式。雨水从屋顶流到立面上,留下了铜绿和苔藓的痕迹,多年来重塑了建筑的特征,不断改变着净水厂的外观。

净水厂的主题是水,在建筑立面上有所体现。喷射混凝土是一种可持续的建筑材料,要在喷射混凝土还干时运送到建筑场地,并在使用前与水混合,这一点非常关键,因为建筑场地处在饮用水保护区内。理想的情况是:粗粒喷射混凝土上能够出现铜绿的光泽。因此,净水厂的立面将融入到环境之中。净水厂与自然进行对话,它要做自然的邻居,而不是做自然的主人。

Beside some woodland, an asymmetric form rises from the ground like an organic outcrop of rock with cave-like openings. This is the new municipal water purification plant by Oppenheim Architecture, commissioned by the City of Muttenz, Switzerland, to provide its drinking water. The plant is a model of sustainability, extremely sensitive to its contrasting setting near the Rhine riverfront. Settled between the protected forest and the nearby industrial parks, the project also incorporates an educational function to display the complex purification process in such a stressed environment.
A unique and important function of the drinking water treatment plant is to create a new landmark for the town of Muttenz and the Basel area. The role of the architecture is to link and express the unique and state-of-the-art technology, placed in a natural ecosystem and emphasizing the importance of the purification process.
The building is located inside a lush green forest next to the River Rhine. This contrasting context – the natural and the industrial – is addressed by the architectural concept. The engineering-driven arrangement of the building's inner life defines its form and the size. Like a tight dress, the skin presses against it and represents the technical inner life to the outside. Pipelines, filters and apparatuses can be read through the facade in an abstract manner. The result is an expressive building, acting like an "objet trouvé" (found object) in its natural context, reduced to its materiality and form.
The drinking water purification plant houses a state of the art, 3-phased treatment process, which will ensure the highest quality of water for the citizen of Muttenz. A public area invites the population to appreciate this complex process and celebrate water with all senses. The desired pedagogical effectiveness is achieved by a circulation route that guides visitors through the different building areas and process phases. An open platform serves as a collection area and as a stage for the presentations.
The alcove-like room is pure, open to the outside and sits on a pool of water, reflecting the daylight and collecting the rainwater pouring in from the roof. Depending on the time and season, it will be moist, cold and mystical. The water will be experienced in all senses. The opposites of the context are playfully implemented in the facade material: Soft in expression, crude and hard in its texture. Shotcrete enforces the expressive facade and is used on all sides and the roof. The shotcrete is treated in a way to allow a porous texture. Rainwater flows from the roof over the facade, leaving a patina and moss stains, re-characterizing the building over the years and constantly changing the appearance of the plant.
The theme of water is represented on the facade. Sprayed concrete is a very sustainable building material, which was delivered dry to the construction site and was mixed with water only before the application. This is an important aspect, because the construction site was in a drinking water protection zone. It is desired that the coarse-grained shotcrete will acquire a green patina. Thus, the facade will merge with its context. The building is in dialogue with nature. It is built with nature, not on top of nature.

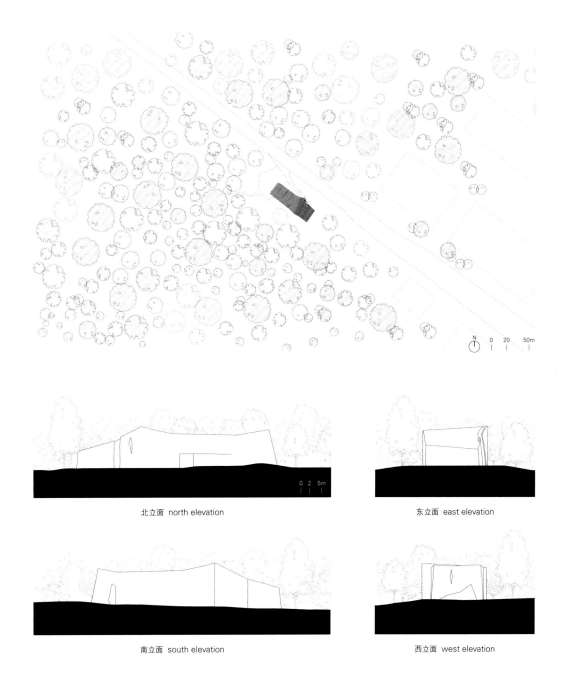

北立面 north elevation

东立面 east elevation

南立面 south elevation

西立面 west elevation

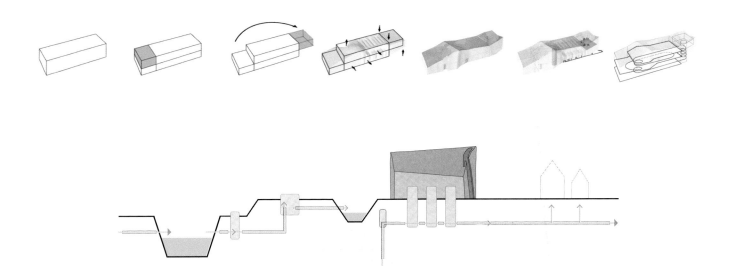

131

项目名称：Muttenz Water Purification Plant / 地点：Muttenz, Switzerland / 建筑师与室内设计师：Oppenheim Architecture / 主管负责人：Chad Oppenheim, Beat Huesler
项目经理：Frederic Borruat / 项目资助方：Aleksandra Melion, Tom Mckeog / 总承包商：ERNE AG / 总体规划：CSD Engineers / 结构顾问：WMM Engineers AG / 喷射混凝土
顾问：Greuter AG / 客户：City of Muttenz / 用地面积：3,000m² / 建筑面积：1,850m² / 竣工时间：2017 / 摄影师：©Borje Müller (courtesy of the architect) - p.126, p.130, p.137, p.138~139, p.141; ©Aaron Kohler (courtesy of the architect) - p.128[lower], p.131, p.132, p.133; ©Beat Huesler (courtesy of the architect) - p.124~125; ©Rasem Kamal (courtesy of the architect) - p.128[upper], p.136, p.140

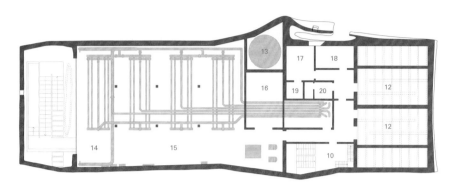

二层 first floor

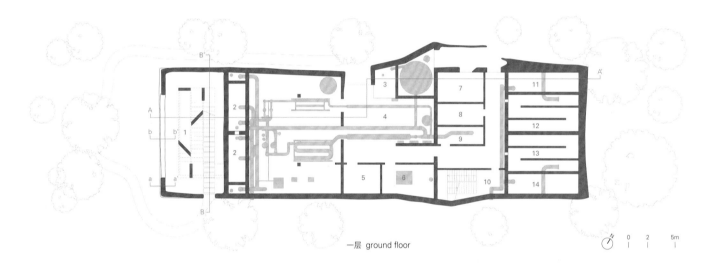

一层 ground floor

1.入口大厅 2.原水储存罐 3.PAK回收/缓冲池 4.交通流线空间 5.化学品储藏室 6.工作坊 7.电气室 8.高压+总开关室 9.机械轴 10.楼梯 11.固定床吸收器 12.PAK接触池 13.PAK地窖 14.空气压缩站 15.超滤设备 16.中央通风装置 17.休息室 18.办公室 19.生物洁净室 20.卫生间
1. entry hall 2. raw water tank 3. PAK-recycling/buffer pool 4. circulation space 5. chemicals storage 6. workshop 7. electrical room 8. HV+MS room 9. mechanical shaft 10. staircase 11. fixed bed absorber 12. PAK contact basin 13. PAK silo 14. compressed air station 15. ultra filtration plant 16. central ventilation 17. lounge 18. office 19. cleaning room 20. W.C.

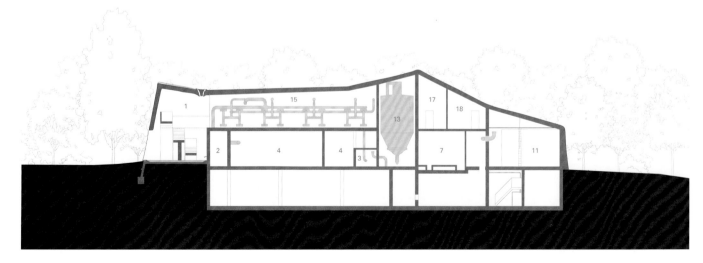

A-A' 剖面图 section A-A'

a-a' 剖面详图 detail a-a'

b-b' 剖面详图 detail b-b'

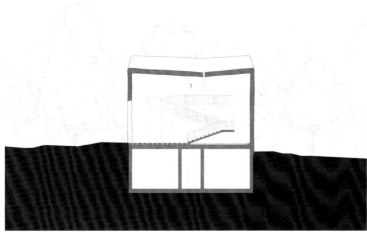

B-B' 剖面图 section B-B'

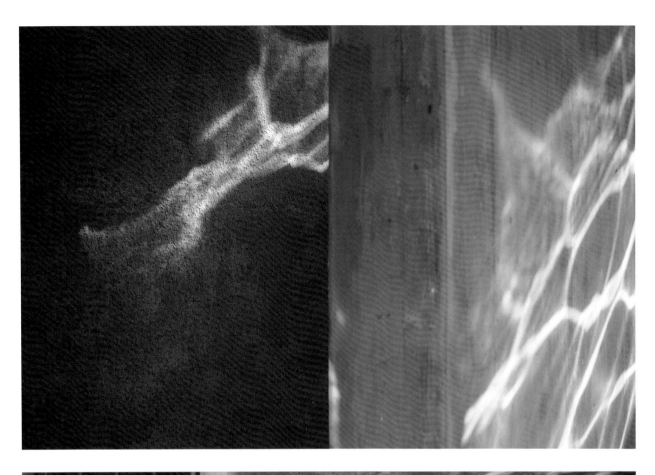

建筑重现

The Palimp

我们不应该畏惧"欧洲中心论",因为欧洲城市同其他城市的一大区别就是:它们对于保护历史建筑和历史文脉的关注程度不同。有些国家对历史的关注可能是过度的。在意大利,相关保护机构经常严禁建筑师对历史建筑内外进行明显改动。本节展示的四个项目表明,历史中心经人为干预后,会成为一种极具创造力的新建筑。这类建筑结合了新与旧,而且新覆盖于旧之上,类似于古老建筑的重现。虽然欧洲人保留中心城市的历史建筑是出于极强的文化和感情原因,但其实这背后也有其潜在的可持续因素。在美国和许多东方国家,大规模拆除陈旧的基础设施会产

We should not be afraid of eurocentrism to note that one of the major differences between European cities and those elsewhere is their attention to preserving historic structures and historic contexts. This attention to the past can be excessive in countries like Italy, where preservation authorities frequently forbid architects from making significant changes, outside or inside of historic buildings. But as shown in the four projects under review, an intervention in a historic center can often lead to a very creative synthesis, where a layering of new upon old resembles the ancient palimpsest. While Europeans hold strong cultural and sentimental reasons to save the historic fabric of center cities, there is also an underlying sustainable factor in doing so. Although there will always be the added expenses of adjustments to antiquated

Z33当代艺术、设计与建筑美术馆_Z33 House for Contemporary Art, Design and Architecture / Francesa Torzo Architetto

雅各比工作室_Jacoby Studios / David Chipperfield Architects

洛桑州立美术馆_Cantonal Museum of Fine Arts / Barozzi Veiga

科隆学校附属建筑和住房_School Annex and Housing, Cologne-Lindenthal / LRO Lederer Ragnarsdóttir Oei Architekten

建筑重现_The Palimpsest Effect / Richard Ingersoll

sest Effect

生大量垃圾废料以及细小颗粒污染。因此，尽管整修陈旧的基础设施会产生额外花销，在新旧结合的混合建筑中小规模进行改造也会耗费更多时间，但建筑重现可以避免美国和东方国家随处可见的这种状况。值得注意的是，这四个项目都是"以人为本"的，而且它们都融入到了当地的环境中，建筑结构松散，令人感觉舒适。建筑师的个人品味蕴藏其间，同时兼顾了历史沉淀，建筑变得新颖却并未抛弃过去。当今的建筑文化存在建造华丽建筑的"风气"，但这四个项目整肃了这种"风气"，即建筑的新旧结合可以达到浑然一体的程度。

infrastructures and an excess of time to apply small-scale craft in the mix of new structures with old, the palimpsest effect avoids the enormous waste of trashed materials and fine particle pollution that come from large demolitions, such as those seen all over the USA and the Orient. The first thing one notices in these four projects is their attention to human scale. They feel comfortable, always helping to loosen up the fabric by creating better access to their surroundings. They are novel without being hostile, expressing their author's taste without ignoring the historic layers. The lesson they offer to an architecture culture that has been overly dominated by iconic solutions, is that they fit so well in their contexts they may go unnoticed.

建筑重现
The Palimpsest Effect

Richard Ingersoll

你是不是经常会听到建筑商坚称"拆毁重建,成本更低"?清除建筑场地重新开始的要求,不仅是勒·柯布西耶和城区改造拥护者做出的倡议,更深深地扎根于建筑行业。为了清除场地,就要消除历史痕迹,这样才能选用新材料,更容易规划规模统一的新基础设施。除了欧洲城市的历史中心,大多数室内和室外项目都是这样做的。因此,在欧洲,我们必须另辟蹊径,摒弃"清除建筑场地重新开始"的心态,就有了所谓的"建筑重现"。

"重新开始"和"建筑重现"的隐喻来自古希腊和古罗马人的学术练习,他们会用覆盖着蜡的古老写字板写字。每当学生想要重新开始,就会在写字板上融化一层新的蜡,叠加出一层新的空白书写页面。但在通常情况下,在新的一页上仍然可以看到之前一页的痕迹,这种留有痕迹的分层法就是所谓的"重写(重现)"。

意大利建筑师卡洛·斯卡帕的作品有翻新意大利维罗那的韦基奥城堡、翻新威尼斯的奎瑞尼·斯坦帕里亚基金会建筑,他似乎很乐意将自己的建筑设计细节融入到现有建筑的细节中,在不破坏历史本质的情况下,创造出新的历史建筑。另一位意大利建筑师马西莫·卡尔马西,完善了一种翻新历史建筑的方法,这种方法就是展示建筑不同的时间层次,但并不是要说明一个时期优于另一个时期。譬如,他最近改造的奥地利军用面包厂,坐落在维罗那,就是一个很典型的例子,现在这家面包厂改造成了大学。也许最激进的建筑重现的实践者是比利时的建筑师事务所ROTOR, D.C. (D.C.代表拆毁与咨询),他们去建筑场地拆卸,抢救地砖和扶手等还能使用的材料,然后在自己的建筑项目中再利用这些材料。谈到"清除建筑场地重新开始"这一问题时,他们表示:"在拆

How often have you heard a builder insist "it will cost much less to tear it down and start over"? The mandate for a clean slate, for a tabula rasa, was not just an agenda promulgated by Le Corbusier and the protagonists of Urban Renewal, but is deeply rooted in the trades. To scrape a site clean, eliminating its historical traces in order to have an easier time to produce a project with uniform scale, new infrastructure, and new materials, is the norm for most situations, indoors and outdoors, excepting the historic centers of European cities. So it is here we must look for an alternative to the mentality of tabula rasa, which can be called the "palimpsest effect". In both cases, the metaphor derives from the ancient writing tablet covered with wax, used by the ancient Greeks and Romans for their scholastic exercises. Each time a student wanted to start over, a new layer of wax was melted on the tablet to create a fresh page of tabula rasa. But it was usually the case that the markings from previous layers could still be seen through the new layer, a stratification of traces known as palimpsest. The Italian architect, Carlo Scarpa, in works such as the renovation of Castelvecchio in Verona and the Fondazione Querini-Stampalia in Venice, seemed happiest mixing his own details with those of existing buildings. He made historic structures new without ruining their historical essence. Another Italian architect, Massimo Carmassi, has perfected a method of renewing historic structures, showing their different layers of time, without privileging one period over another, such as his most recent work, the Austrian Military Bread Ovens in Verona, now converted into the university. Perhaps the most radical practitioners of palimpsestism are the Belgian architects ROTOR, D.C. (the D.C. stands for Deconstruction and Consulting), who go to demolition sites to salvage materials, such as floor tiles and handrails, and then recycle them in their projects. Against tabula rasa they claim: "In demolishing,

卡斯特维奇博物馆，意大利维罗那
Castelvecchio Museum, Verona, Italy

毁建筑时，开发商没有考虑到这么做会对社会产生什么影响。"

科隆一所学校的综合体（204页）的扩建由LRO（Lederer Ragnarsdóttir Oei Architekten）建筑师事务所设计，园区位于郊区，远离科隆繁华的街市，这是四个项目中唯一没有建设在历史色彩浓厚的街道中的建筑。教育园区的右侧是一座基督复活教堂，由获奖的德国建筑师哥特佛伊德·波姆设计，采用了新表现主义的手法。在原有教堂建筑的立面上采用了砖块和混凝土的混合结构，建筑师为音乐厅、教室、宿舍和自助餐厅设计了一个相当紧凑的镶砖结构，这些设施也对园区现有三所学校的音乐家和唱诗班成员开放。尽管没有按照以前的建筑结构改造，但通过宽敞的拱门和偶尔能看到的曲线，还是可以感觉到与历史建筑的共同点的。扩建后的新楼规模不大，看起来不像是典型的学校，它巧妙地将现有教室的社会空间聚集在一起，这些教室以不规则多边形建筑的形式散布在教堂庭院的周围。

第二个项目是雅各比工作室（170页），位于德国一座中等大小的城市——帕德伯恩，该项目称得上是大卫·切波菲尔德建筑师事务所最精致的建筑重现作品。在中世纪，雅各比工作室位于市中心，当时这里是一座卡布奇修道院，还有一座小教堂，经历了数次重建改造，现在用作会议厅。该修道院的一部分在第二次世界大战期间遭到摧毁，后来改建为医院，并于2003年迁往别处。大卫·切波菲尔德建筑师事务所当时刚刚完成柏林新博物馆的翻新任务，拆除了场地不规则的扩建部分，留下了展示17世纪风貌的光秃秃墙壁，同时，采用垂直设计

developers do not consider the true cost to society".
The expansion of a school complex in Cologne (p.204) by LRO Lederer Ragnarsdóttir Oei Architekten, built in the outer ring of Cologne's concentric urbanization, is the only one of the four projects not set in a dense network of historic streets. The neo-Expressionist church of the Resurrection of Christ, designed by Prize-winning German architect Gottfried Böhm, is located right next to the school complex. Following the mix of brick and concrete on the facades of the original church structure, they have added a fairly compact brick-clad nucleus for concert rooms, classrooms, dorms, and a cafeteria, accessible to the existing three schools for musicians and choir singers. The broad arches and occasional curving reveals allude to the original building without copying it. Without appearing institutional the new building maintains a low scale and deftly gathers in the social spaces of the existing classroom clusters that spread in informal polygons around the churchyard.
The second project in Germany in the medium-size city of Paderborn, the Jacoby Studios (p.170), provides the most exquisite palimpsest of the group. Situated in the city's medieval center, the complex was once a Capuccin monastery with a small church, now used as a conference hall, that went through several permutations. The convent was partly destroyed during WWII and then retrofitted as a hospital that moved elsewhere in 2003. David Chipperfield's Berlin office, fresh from the renovation of the Neues Museum in Berlin, cleaned the site of incongruous additions and left the bare walls from the 17th-century to display their mineral materiality, while adding new wings in a pinwheel order, employing the discreet, orthogonal vocabulary found in many of their best works. Chipperfield's architecture, while consciously avoiding formal originality, conveys a unique, recognizable sensibility,

科隆学校附属建筑和住房,德国科隆-林登塔尔
School Annex and Housing, Cologne-Lindenthal, Germany

的方式,按照纸风车的顺序为建筑物增加新的翼楼。大卫·切波菲尔德建筑师事务所的许多杰作都采用了这样的设计方式。他们的建筑设计不仅刻意摆脱了形式上的独创性,还赋予了建筑一种引人注目的独特风格,和环境十分兼容,给人一种十分舒服的感觉,就像传统风格的家具一样。新空间的通透性极强,每个开间都安装了跟开间跨度差不多一致的全长通高玻璃窗,与原有建筑厚重的围墙和老旧通道形成了鲜明的对比。设计师利用当地的溪流,启动热泵,以可持续的方式进行供暖和制冷,这是利用当地环境来改善建筑的又一典范。

今年意大利建筑奖得主,年轻的意大利建筑师弗朗西斯卡·托尔佐,在比利时小镇哈塞尔特的Z33当代艺术、设计与建筑美术馆(148页)的扩建中,更加注重规模、类型、颜色和选材。当然,其中也有拆除的部分,包括一扇18世纪的大门,但新建的翼楼外部贴着葡萄酒色的陶土砖,与拆除的部分相比,新建部分让人感觉更适合这片曾经是修道院的场地。方砖的砌法与古罗马的方锥形石块饰面类似,为带有两处优雅曲面的整面混凝土墙增添了神秘的光泽。建筑师将方砖锯齿状的边缘裸露出来,似乎是在说明砖只是"外衣",这堵墙并不是用砖头砌成的。博物馆原有的翼楼中曾经有一所学校,在主楼与翼楼的结合处,建筑师设计了一个又高又窄的缺口,两层楼高,让人可以窥见一个神秘的庭院入口,庭院内部豁然开朗,而从庭院外部看上去则封闭又黑暗。新建翼楼内部每个房间的大小都不一样,而且在走动时,给人以不同的光感和空间感。建筑内部所有门槛和开窗洞口内都设计了深深的斜角门窗框,看起来像是孔隙,而不像是门窗的形状。建筑师在天花板的设计上采用了泡状混

keenly sensitive to the context, that creates a fitness like good Biedermeier furniture. In contrast to the solid enclosures of the original cloister and the historic passages, the new spaces are startlingly transparent, each bay with full length windows. That the local stream was channeled to create a thermal pump for sustainable heating and cooling is just another aspect of improving while using the layers of the context.

The young Italian architect, Francesca Torzo, this year's winner of the Italian Prize for Architecture, paid even greater attention to scale, type, color, and materials in her addition to the Z33 House for Contemporary Art, Design and Architecture (p.148) in the small Belgian town of Hasselt. Demolitions occurred, including the removal of an 18th century gateway, but the new wing, clad in wine-colored terracotta bricks, offers a much better sense of fit into the enclosure that once was a beguinage, than what was removed. The square bricks, arranged like ancient Roman opus reticulatum, give an uncanny patina to a monolithic concrete wall, gracefully bent in two places. The architect has left the sawtooth edges of the bricks exposed as if to indicate that they are cladding and not structural. At the juncture with the existing wing of the museum, which had once housed a school, she cut a tall narrow gap, two stories high, which invites one to peer into a mysterious entry court, as open and luminous as the exterior is closed and dark. Each of the rooms in the new wing is different in size and geared to creating a different sensation of light and scale as one moves about. All of the interior openings for thresholds and fenestration have been framed with deep bevels, making them read as apertures rather than doors or windows. The architect has designed the ceilings as quilted concrete, softening their presence. On the courtyard side of the new wing, she has put long narrow windows, similar to

洛桑州立美术馆，瑞士洛桑
Cantonal Museum of Fine Arts, Lausanne, Switzerland

凝土，使其看起来柔和一些。她在新建翼楼的庭院一侧安装了狭长的窗户，与相邻的翼楼相似。这里过去曾是一家杜松子酒厂，她将新建翼楼的立面与原先工厂的斜屋顶相连。建筑整体看起来像是新的一页，引人驻足欣赏，也揭示了相邻历史建筑的本质。

值得注意的是，在洛桑州立美术馆（188页）这个已经竣工的项目中，也出现了一定程度的"场地清除"，因为破旧的火车棚已经完全被夷为平地。但巴罗齐·维加建筑师事务所倾注了心血来维持建筑场地的规模，并将其中一栋被拆毁建筑的一部分合并到空置的混凝土立面中，混凝土立面朝着南面的铁路。建筑师们是新一代的形式主义者，他们信奉不妥协的欧几里得几何，例如，洛桑州立美术馆就是一个长方形的平行六面体。但建筑师们以一种建筑重现的姿态，将一个拆毁建筑的金属轮廓嵌入短立面，人们在进入美术馆时，一眼便能看到这个立面。建筑师们在长长的东北立面上增加了一排一米宽的混凝土翼板，在它们中间只插入了一个作为入口的突出的混凝土立方体。走进这个设计简朴的立面，便是一个巨大的、带天窗的三层大厅，让人想起19世纪装潢宏伟的车站。走廊尽头的楼梯在大拱顶下方向上延伸，与前文提到的历史碎片相契合。建筑师在上层设计了一排而立的画廊，这是一系列带有采光天窗的白色盒子，天窗嵌入深深的拱腹中。与之前介绍的项目一样，洛桑州立美术博物馆还是有一定程度的拆除，但必须保持对城市环境现有类型、规模、材料和颜色的尊重，既体现出老城的饱经风霜，又令人耳目一新。

the neighboring wing and hooked the elevation into the pitched roof of what was once a gin factory on the far end. The whole project appears like a fascinating new layer that reveals the essence of the adjacent historic structures.

One should notice that a certain tabula rasa occurred at the concluding project, the Cantonal Museum of Fine Arts in Lausanne (p.188), since the dilapidated train sheds were completely razed. The architects, Barozzi Veiga, however, put great effort into maintaining the scale of the site and incorporated a piece of one of the demolished buildings into the otherwise vacant concrete facade that looks toward the train tracks on the south. The architects belong to a new generation of formalists, pursuing uncompromised Euclidian geometry, in this case an oblong parallelepiped. But as a gesture to palimpsest they applied a metal profile of one of the demolished structures on the short facade that one first sees while entering the site. They added to the long northeast elevation a row of meter-deep brick fins that are only interrupted by a single protruding concrete cube that indicates the entry. One enters the purposely humble facade to find an immense, skylit, three-story hall that recalls the grandeur of 19th-century stations. The stairs at the end of the axis climb under a grand vault fit into the historic fragment mentioned above. The architects arranged the upper galleries enfilade, a series of white boxes served by skylights set into deep soffits. As in all of the featured projects the palimpsestism of the new museum in Lausanne does not preclude demolitions, but necessarily maintains respect for an urban setting's existing typologies, scale, materials and colors. It feels refreshingly new while alluding to the seasoned character of the old city.

Z33当代艺术、设计与建筑美术馆
Z33 House for Contemporary Art, Design and Architecture

Francesca Torzo Architetto

具有历史意义的城市结构与抽象的美术馆扩建
Historic urban fabric informs an abstracted museum extension

比利时哈塞尔特市的Z33美术馆与各式各样的砖砌建筑共同组成了一个三角形，三角形的中间是一座大型中央花园，这些砖砌建筑是为了还没有成为修女的信奉宗教女性而建的。这种房子的排列组合也被称为女修道院，最初是为了将城市排除在花园之外，而花园是这些女性的公共资源。如今，位于比利时哈塞尔特市的这一建筑群囊括了Z33美术馆，该馆的老楼从1958年使用到现在，在2019年时完成了4664m²的扩建，由意大利建筑师弗朗西斯卡·托尔佐设计。新建部分重新解读了旧建筑表达出的感觉，其建筑外形也和当地的大环境产生了呼应。该设计营造出了一方休憩之所，园中寂静也与市中心的喧嚣对比鲜明。Z33美术馆附近有一家建于19世纪的杜松子酒厂，现在已成为一个美术馆，与Z33美术馆共同在公园中营造出一种舒适之感，透过一扇扇嵌入墙壁凹处的窗子便可一探究竟。

美术馆的外立面以一种封闭的状态临街而立。外立面长60m，高12m，是一堵长长的、无接缝的实体砖墙，几乎没有开口，沿着街道的边界线有一个轻微的垂直转折。手工制作的34 494块砖颜色和大小各不相同，一个接一个地摆放，折弯处只留有一个膨胀接缝。砖砌的图案呈菱形。墙面所使用的砖块看上去老旧粗糙，还带有些铜绿，但却与这里的环境相衬。临街的围墙使美术馆和花园的宁静不受打扰。这堵墙也成了城市生活的背景板，映照了树木、城市灯光和行人的影子。

通过外墙上高高的长方形洞口，便能找到入口。里面是一个小天井，天井里种着一棵树，越过这里，通过前厅可以进入1958年的翼楼和新展厅。展厅一层的地势比街道要高出一米，但该区域同街道、花园完美融合。一条三层高的白色连接走廊位于临街围墙之后，到拐弯处逐渐变宽。扩建部分中央的内部庭院将光线引入室内空间。

与1958年的美术馆相同，扩建部分有两层公共空间，但与房间的传统基础设施（即比例十分精确，各个房间都有一定的隐匿性）形成鲜明对比的是，扩建部分包括了大小、比例、光线和氛围各不相同的展厅，风格简约。在一间展厅可以看到另一间展厅，提供了各种各样的视角，可以看到不同的风景。各个展厅之间的门槛也让人意识到这些空间是依序分布的，在穿过该建筑时，给人一种时间感和深度感，让人意识到空间的尺度是不同的。这些门槛位于展厅入口处以及展厅与展厅之间像窗户一样的内部洞口，展厅的曝光度和私密性是由门槛的比例调节的。由于这些门槛的框架加宽了倾斜梯形边的墙壁，也降低了墙壁的存在感，因此门槛本身的存在也更加抽象。

The art museum Z33 in Hasselt, Belgium forms part of a triangle of varied brick buildings enclosing a large central garden, which were built for independent religious women who did not become nuns. This arrangement is known as a beguinage, and originally served to exclude the city from the garden, which was a communal resource for the women. Now, the buildings of the Hasselt Beguinage include the Z33 museum, comprising an existing building from 1958, and a 4,664m² extension completed in 2019 and designed by Italian architect Francesca Torzo. The new building reinterprets the expression of the old buildings, and its form responds to the context. The design creates a place of rest, with a different sound from the bustling of the city center. Together with a 19th century neighboring gin factory, now a museum, it creates a niche in the garden, which it looks out to through numerous recessed windows.

The exterior facade chooses to be silent towards the street. It is a long jointless solid brick wall, 60m long and 12m high, with few openings and a shallowly angled vertical fold following the street line. 34,494 handmade bricks with variations in color and size were placed one by one, with only one dilatation joint at the folding. A diamond pattern characterizes the brickwork. Rumble bricks, which have patinas and look worn, give Z33 its response to context. The street wall protects the quietness of the art museum and the garden. It also acts as a background for the life of the city, its trees and urban lights and the shadows of passing pedestrians.

The entrance is through a tall rectangular opening in the exterior wall. Inside is a small patio in which a tree grows. Beyond it, the vestibule gives access to the 1958 wing and to the new rooms. The area blends the street level with the exhibition ground floor, which is raised by one meter, and the garden level. A triple-height connecting corridor rendered in white lies behind the street wall and follows its bend but widens in width. An internal courtyard in the center of the extension brings light into interior spaces.

Like the 1958 museum, the extension has two public levels, but in contrast with its classical infrastructure of rooms with fine proportions and a certain degree of anonymity, the extension is an ensemble of simple rooms that vary in size, proportion, light and atmosphere. The rooms look into each other, offering multiple perspectives and views. The thresholds between spaces allow recognition of the spatial sequences, give a sense of time and depth to the routes through the building, and offer a perception of the different scales of the spaces. Exposure and intimacy are regulated by the proportions of the thresholds, which include room entrances and window-like internal openings between spaces. Because the threshold frames widen out towards the walls with angled trapezoidal edges, they dematerialize the wall, making the thresholds themselves more abstract.

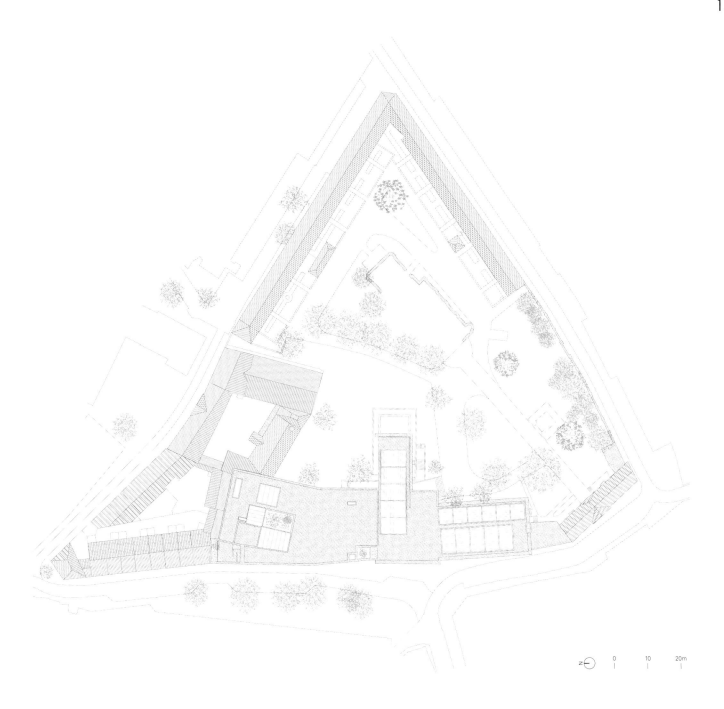

东立面 east elevation

西立面 west elevation

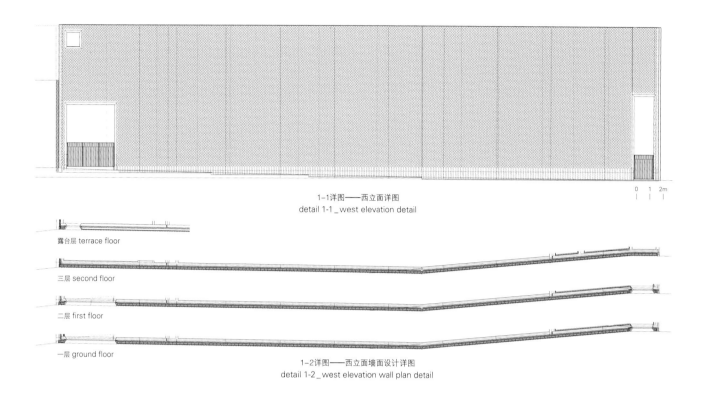

1-1详图——西立面详图
detail 1-1 _ west elevation detail

1-2详图——西立面墙面设计详图
detail 1-2 _ west elevation wall plan detail

露台层 terrace floor
三层 second floor
二层 first floor
一层 ground floor

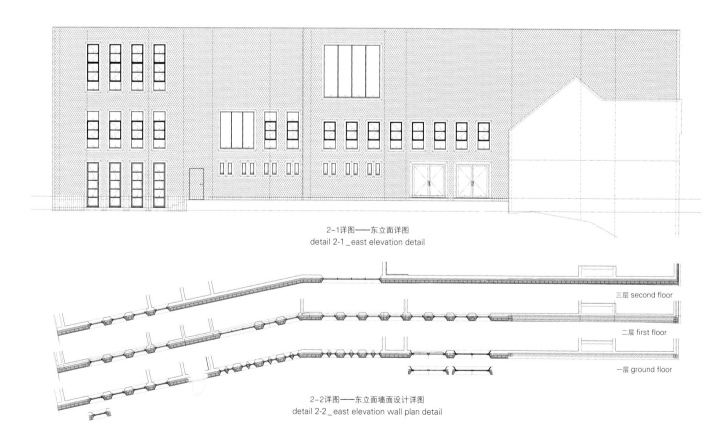

2-1详图——东立面详图
detail 2-1 _ east elevation detail

2-2详图——东立面墙面设计详图
detail 2-2 _ east elevation wall plan detail

三层 second floor
二层 first floor
一层 ground floor

采访弗朗西斯卡·托尔佐与赫伯特·赖特
Interview_Francesca Torzo + Herbert Wright

问:当您在构想Z33美术馆的建筑理念时,您有没有参考过女修道院的历史和住在这里的女教徒的生活?
答:我认为这个女修道院不仅仅关乎宗教,更是一个充满特色的文化场所。尽管表现方式有所不同,但扩建的建筑正好延续了城市的风姿和特征。

问:临街围墙不仅是将建筑与城市隔绝的防护屏障,也是描绘城市光影的画布。建筑师是否应该更多地考虑这些无窗墙壁对城市氛围的潜在作用?
答:我觉得应该换一种说法。临街立面更像是城市"内部"的保护伞,与城市共同合作。我认为一般建筑师可以提出,也应该提出更多的问题。

问:美术馆的内部空间就像一个三维的、相互连接的矩阵,随着观展者的走动,光线和空间感会慢慢转变。为什么没有采用传统画廊或当代画廊常见的"白立方体"美学,而采用这种方法?
答:这座建筑建于1958年,采用典型的展厅排序,在过去十年中,一直是理想的白色立方体展览空间。但设计方案的目的是提供一些前所未有的内容。事实上,Z33美术馆现在由两部分组成。文化部分借鉴于一场谈话,谈话双方是扬·荷特(1936—2014年)与哈洛德·塞曼(1933—2005年),扬·荷特长期以来一直担任比利时根特SMAK当代美术馆的策展人,而哈洛德·塞曼是瑞士策展人,其革命性的方法重新定义了国际艺术策展。

问:您是怎样让这座建于1958年的建筑始终如一,并和它保持羁绊的?
答:凭借我的记忆、决策和脚踏实地的反思。

问:楼梯栏杆采用独特的钢结构,用于固定钢结构的斜角钢支撑用圆盘状物固定。您能告诉我们这个设计背后的想法吗?
答:这是对1958年建筑栏杆的呼应;这个设计提炼了其表现形式,没有任何模仿。我们对效果也感到非常满意。

问:Z33美术馆中的一些想法(例如,墙壁、门槛)是否借鉴了您早期的作品,例如,您2010年的索拉诺Due住宅?
答:每个项目都有特定的文化背景,都在讲述不同的故事。Z33美术馆是公园周围的围墙的一部分,其设计目标是按照传统的砖砌建造方式来建造其立面。走在街上,人们会感受到一面长长的实心砖墙带来的宁静之感。扩建建筑囊括了各种各样朴素的展厅:空间格局的复杂性也呼应了城市体验的多样性,反映了公共空间和私人空间、暴露区域与私密区域之间是有梯度变化的。

问:有没有其他的项目或地方,无论是最近的还是过去的,激发了您的灵感?
答:我脑海中存在的一切就是我灵感的源泉。

Did you think about the history of the beguinage and the lives of the religious women who lived there when you were developing the architectural concept of Z33?
The beguinage is a cultural place with specific characteristics, beyond religious subjects. The extension building is just in continuity with its urban posture and character, though with a different expression.

The street wall is a protective barrier against the city, but also a canvas for shadows and light from the city. Should architects generally think more about the atmospheric potential of blind walls?
I would rephrase. The street facade is a protection "within" the city. It collaborates with the city. I think in general architects could and should develop more questions.

The internal spaces are like a three-dimensional interconnecting matrix, where light and the sense of space are gently transformed as you go through. Why this approach, in contrast to classical galleries or the contemporary "white cube" aesthetic?
The existing building from 1958 is a classical sequence of rooms, which has been used in the last ten years as ideal "white cube": The purpose of the design proposal was to offer something different from what they already had. In fact Z33 is now a building made of two parts. The cultural reference is a dialogue between Jan Hoet [1936-2014, long-standing curator of contemporary art gallery SMAK in Ghent, Belgium] and Harald Szeemann [1933-2005, Swiss curator whose revolutionary approach redefined international art curation].

How did you approach continuity and connection with the 1958 building?
Through memory, decision making and pragmatic reflections.

The stair bannisters have a distinctive steel structure with angled steel supports secured by circles. Can you tell us the thinking behind this design?
It is an echo of the balustrade of the building from 1958; the design distilled its expression, without any mime of figures. We are content with the result.

Do some ideas in Z33 (for example walls, thresholds) come from your earlier works, such as the "Casa Due, Sorano (2010)"?
Each project is specific to a cultural context and to a narrative. Z33 is part of this wall of buildings around the park. The design sets the goal of a solid facade in line with the traditional masonry. Walking along the street, people experience the quietness of a long solid brick wall. The extension building is an ensemble of simple rooms that vary: the complexity of the spatial pattern echoes the multiplicity of experiences of a city, with gradients between public and private, exposed or intimate.

Did any other projects or places, recent or old, inspire you?
Everything that stays alive in my memory.

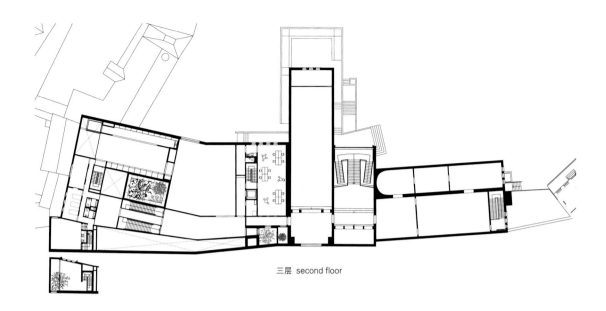

三层 second floor

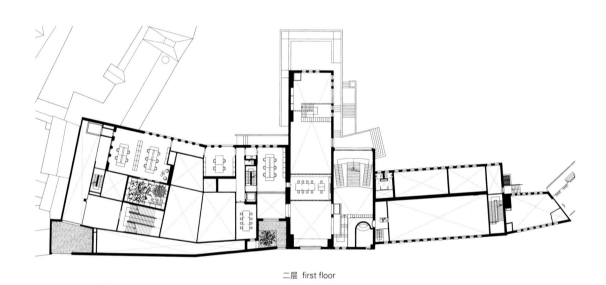

二层 first floor

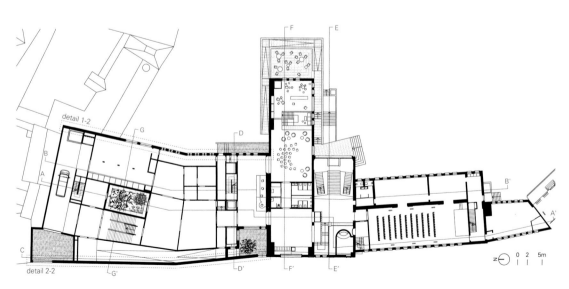

一层 ground floor

项目名称：Z33, House for Contemporary Art, Design and Architecture / 地点：Hasselt, Limburg, Belgium / 建筑师：Francesca Torzo Architetto / 设计团队：Marco Guerra (senior collaborator); Antoine Lebot, Liaohui Guo, Pablo Brenas, Anna Opitz, Riccardo Amarri, Lorenzo Gatta, Elöd Zoltan Golicza, Cyril Kamber, Besart Krasniqi, Jovan Minic, Andrea Nardi, Anna Oliva, Costanza Passuello, Alessandro Pecci, Domenico Singha Pedroli, Nicola Torniamenti, Gion Balthasar von Albertini / 结构工程（结构稳定性顾问——立面结构和镂空天花板）：Gianfranco Bronzini, Conzett Bronzini Partner; ABT België 电气工程：expo light advisor Ben Boving, Gattoni Piazza / 机械工程：Gattoni Piazza / 现场监理：Francesca Torzo Architetto / 地方行政管理：ABT België 其他顾问：Petersen Tegl, Bekaert, Reynaers, Knauf / 承包商：THV Houben Belemco / 客户：Provincie Limburg+Z33 / 用途：Art gallery / 总建筑面积：4,664m²(program), 300m²(exterior) / 竞赛时间：2011.8—2012.8 / 竣工时间：2019.11 / 摄影师：©Gion von Albertini-p.148~149, p.152, p.156, p.157, p.159, p.160, p.161, p.162, p.164, p.165, p.168 ©Courtesy of the architect-p.166, p.167, p.169

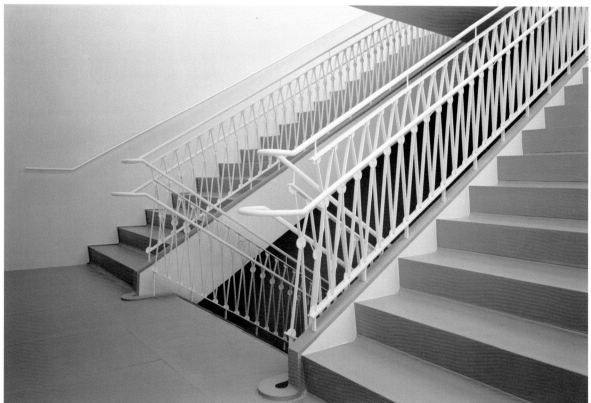

A-A' 剖面图 section A-A' 0 2 5m

B-B' 剖面图 section B-B'

C-C' 剖面图 section C-C'   D-D' 剖面图 section D-D'

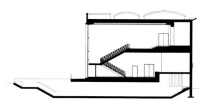

E-E' 剖面图 section E-E'

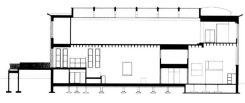

F-F' 剖面图 section F-F'

G-G' 剖面图 section G-G'

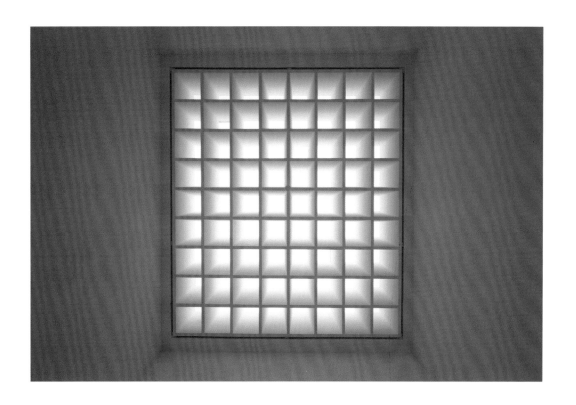

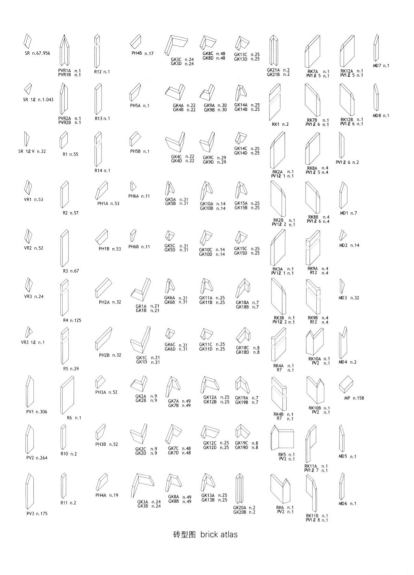

砖型图 brick atlas

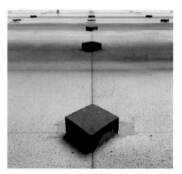

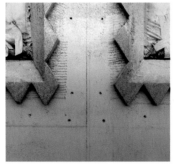

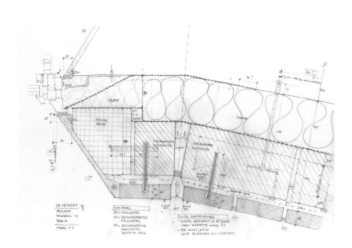

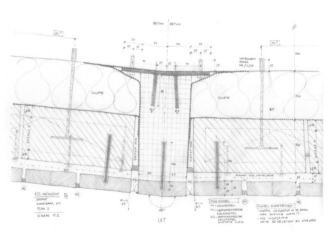

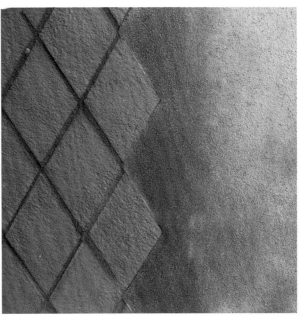

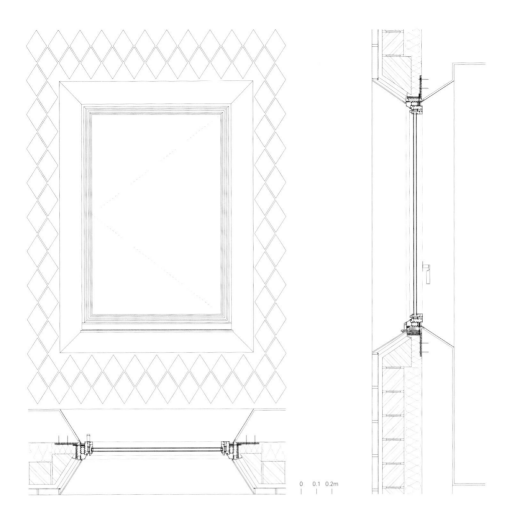

雅各比工作室
Jacoby Studios

David Chipperfield Architects

废墟、古老的立面和新建的翼楼构成了一个办公综合体
Ruins, ancient facades and new wings form an office complex

在德国的帕德伯恩市，大卫·切波菲尔德建筑师事务所改造了一座历史建筑，并将其扩建成为雅各比工作室的总部。原先的建筑在17世纪时是修道院建筑群，从19世纪到2013年是一家医院。

这座新开发的建筑有一个特定的建筑理念，其出发点是建筑场地的历史建筑元素，而工作室决定保留这些元素。同时，该项目还包括：移除战后扩建的部分，将隐藏在重建医院内的历史建筑结构显现出来，一直以来，有些人以为其中一些结构已经丢失。砖石裸露在外，得到修复，或者在需要的时候被完全覆盖，形成了一个风景如画的废墟结构。废墟结构由毛石砌体构成，其核心部分的前身是修道院。教堂的立面和东侧翼楼，以及原来的地下室也保留了下来。教堂的立面上还保留着1659年的开创日期，现在作为主要入口。教堂的内部现被改造成了一个外部空间，即教堂庭院。之前修道院的遗迹也被整合到这个中央庭院中。参观者穿过历史气息浓厚的大门和原教堂，便可进入入口门厅。

新建筑拓展了现有的结构，进一步利用裸露的混凝土和木材开发了整体建筑。新扩建的翼楼高度为二层到三层，位于原建筑的北面、西面和南面。它们的排列顺序与修道院的正交几何顺序一致。

由于保留了大量的外部砖石结构，新建的建筑群再次出现在城市景观中，尽管各个部分大小不一，但整体平衡感十足。在具有中世纪城镇布局的城市景观中，该建筑延续了这一历史风格，但其相邻建筑的主要风格还是20世纪下半叶以来特有的简单立面。新建筑群有着自己的特色，继往开来，镶嵌在由维尔茨国际景观设计公司 (Wirtz International) 设计的公园中。

雅各比工作室在现有的结构中，结合了可持续建筑概念和低技术方法。通过保留现有的建筑结构，使用简单、易维护和耐用的技术，节约了资源，避免了浪费。经过仔细考虑，决定在室内使用机械装置来控制气温，从而取代了以前能耗高的空调系统。附近帕德尔河支流的水通过热泵的方式发电。夏季使用混凝土天花板，用以降温；而在冬季，地板供暖系统负责供热。这两个系统全年都利用了温度几乎保持恒定的河水。

In the German city of Paderborn, the Berlin studio of David Chipperfield Architects has converted and extended a historic building into a company headquarters for Jacoby Studios. The former monastery complex dates from the 17th century. It was a hospital from 19th century until 2013.

The starting point for the architectural concept of the newly developed ensemble was the historical building elements on the site, and it was agreed to protect them. The project involved the removal of the post-war additions, exposing the historic building fabric previously hidden within the rebuilt hospital, some of which had been considered lost. Stone masonry was exposed, repaired and, where necessary, completed, creating a picturesque ruin structure made of quarry-stone masonry with the former cloister at its heart. The facades of the chapel and the eastern wing of the building as well as the original cellars have also been preserved. The chapel facade, which displays its 1659 inauguration date, now serves as the main entrance. The interior of the former chapel was trans-

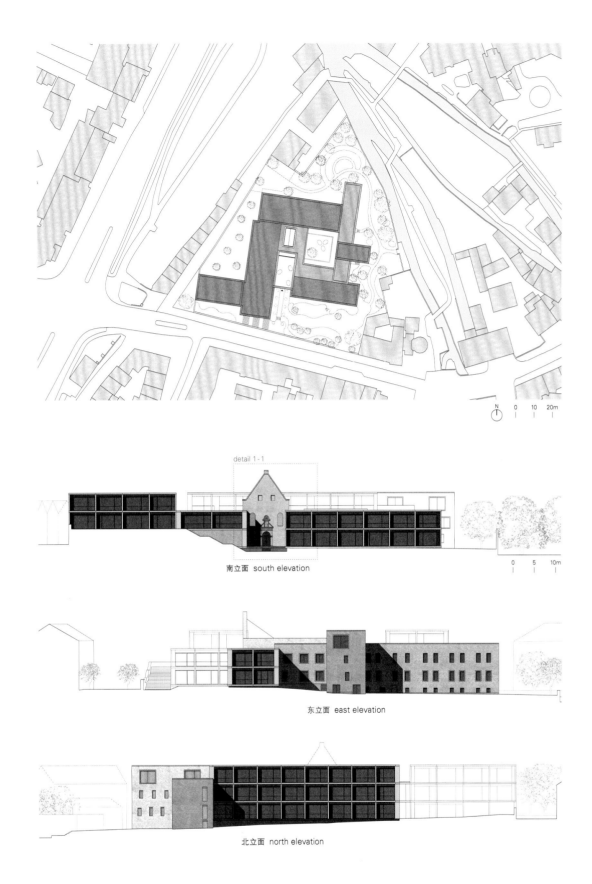

南立面 south elevation

东立面 east elevation

北立面 north elevation

项目名称：Jacoby Studios / 地点：Kisau 8, 33098 Paderborn, Germany / 建筑师：David Chipperfield Architects Berlin / 合伙人：David Chipperfield, Martin Reichert, Alexander Schwarz (Design lead) / 项目建筑师：Franziska Rusch (Concept study), Frithjof Kahl (Preparation and brief to Developed Design, Design intent details, Site design supervision) / 项目团队：Thomas Benk, Thea Cheret, Dirk Gschwind, Elsa Pandozi, Franziska Rusch, Diana Schaffrannek, Eva-Maria Stadelmann, Amelie Wegner; Graphics & Visualisation: Dalia Liksaite / 施工图：Schilling Architekten / 执行建筑师：Jochem Vieren (Project management), Michael Zinnkann (Construction management) / 景观设计师：Wirtz International nv, Peter Wirtz, Jan Grauwels / 结构工程师：Gantert + Wiemeler Ingenieurplanung / 设备工程师和照明顾问：Köster Planung GmbH / 建筑物理与声学：Hansen Ingenieure / 消防顾问：HHP West Beratende Ingenieure GmbH / 客户：Jacoby GbR / 功能：office, staff canteen, photography studio, showroom / 用地面积：8,800m² / 总建筑面积：12,500m² / 楼层：2~3 above ground, 1 below ground / 高度：11.8m above ground / 设计时间：2014 / 施工时间：2017—2020 / 竣工时间：2020 / 摄影师：©Simon Menges

176

重建前，2014年
prior to construction, 2014

历史结构裸露在外的主要立面，2017年
historic fabric exposed-main facade, 2017

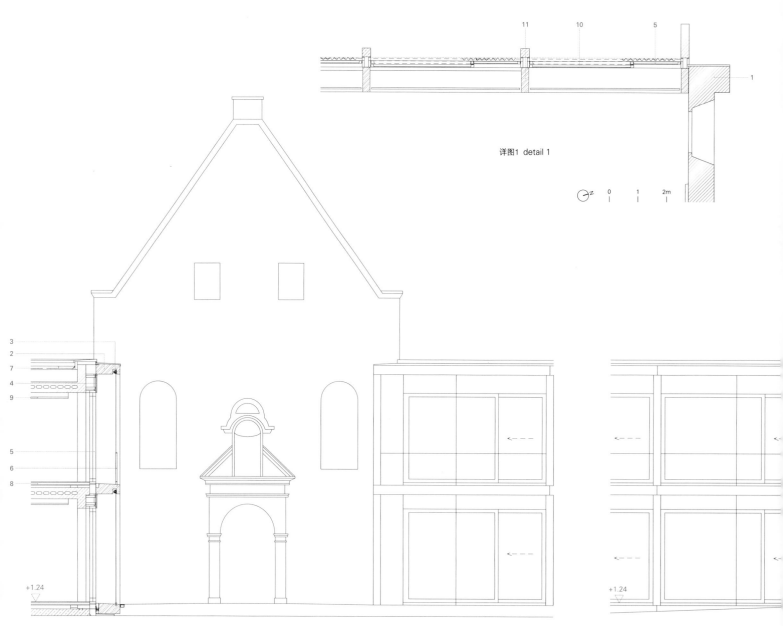

详图1 detail 1

详图1-1 detail 1-1

详图1-2 detail 1-2

历史结构裸露在外的内部庭院, 2017年
historic fabric exposed – internal courtyard, 2017

历史结构裸露在外的东侧翼楼, 2017年
historic fabric exposed – east wing, 2017

东侧翼楼建成以后, 2017年
after completion – east wing, 2017

1. 砖石结构（现存）
2. 预制混凝土构件，横向
3. 外部遮阳设备，垂直遮光帘
4. 外墙板，带橡板饰面的层压木板
5. 木制电梯和推拉门，涂漆橡木
6. 栏杆，扁钢
7. 绿色屋顶
8. 楼层构造，木地板，橡木
9. 照明设备
10. 固定装配玻璃
11. 预制混凝土构件，纵向

1. masonry (existing)
2. precast concrete unit, horizontal
3. external solar shading, vertical roller blind
4. facade panel, laminated wood with oak veneer
5. timber lift and slide door, varnished oak
6. balustrade, flat steel
7. green roof
8. floor construction, plank flooring, oak
9. lighting
10. fixed glazing
11. precast concrete unit, vertical

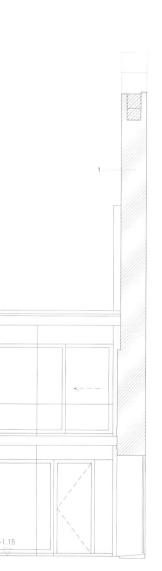
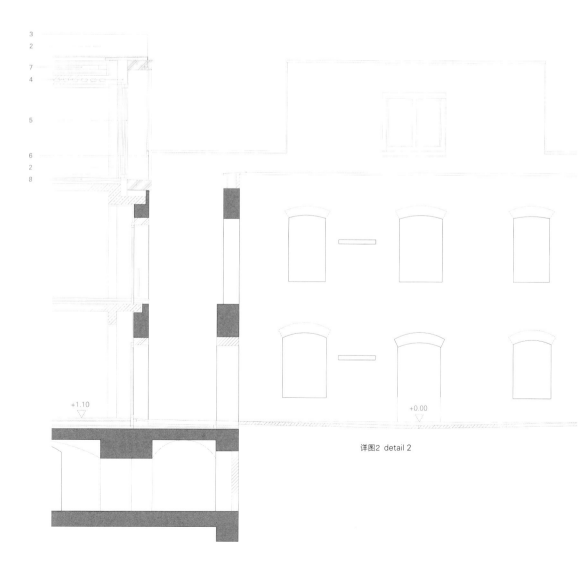

详图2 detail 2

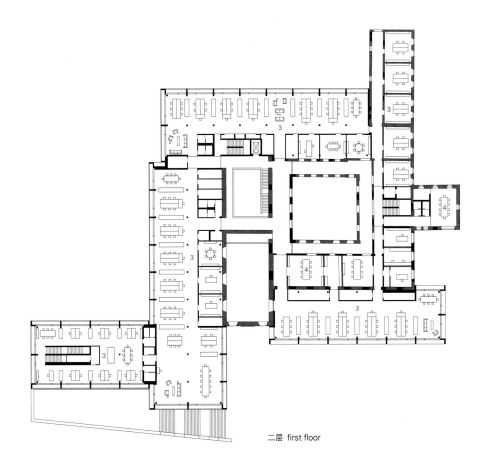

二层 first floor

1. 入口庭院
2. 内部庭院，修道院
3. 办公空间
4. 会议室
5. 餐厅
6. 工作室，展厅
7. 照相馆

1. entry courtyard
2. inner courtyard, cloister
3. office space
4. conference rooms
5. canteen
6. studio, show room
7. photo studio

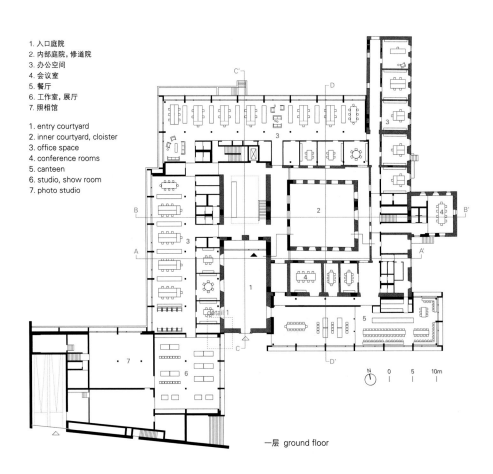

一层 ground floor

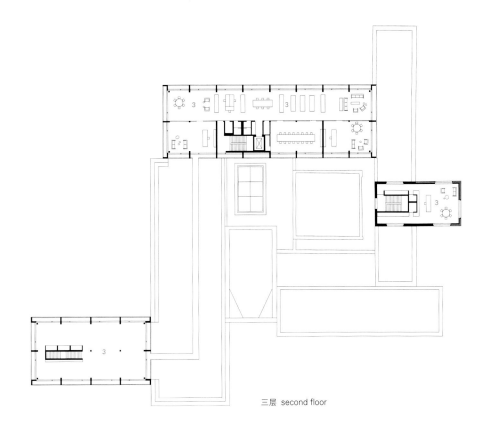

三层 second floor

formed into an exterior space, the chapel courtyard. Remains of the former cloister are integrated into this central courtyard. Through the historic entrance gate and the former chapel, visitors access the entrance foyer.

Extending this existing structure, the new building further develops the ensemble using exposed concrete and timber. New extension wings, ranging from two to three stories are sited to the north, west and south of the original buildings and their arrangement is determined by the orthogonal geometric order of the cloister.

With the extensive preservation of the outer masonry, the new ensemble appears once again in the cityscape as a well-balanced composition of different volumes. It exhibits a high degree of historical continuity within an urban landscape that has a medieval town layout, but where the adjacent buildings are predominantly characterized by simple facades from the second half of the 20th century. The new complex has its own sense of identity, while maintaining historic continuity. It is embedded in a garden designed by Wirtz International.

Jacoby Studios combines a sustainable concept of building within existing fabric with a low-tech approach. By preserving the existing building fabric and using simple, low-maintenance and durable technology, resources were saved and waste avoided. The carefully considered use of mechanical devices for climate control in the interior replaces the previous air-conditioning system which had high energy consumption. Water from the neighboring tributary of the river Pader is used to generate energy by means of a heat pump. For cooling in summer, the concrete ceilings are activated, while in winter, heat is supplied via an underfloor heating system. Both systems use the almost constant temperature of the river water throughout the year.

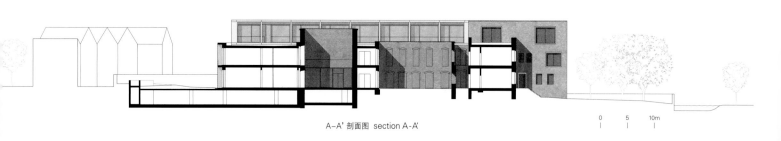

A-A' 剖面图 section A-A'

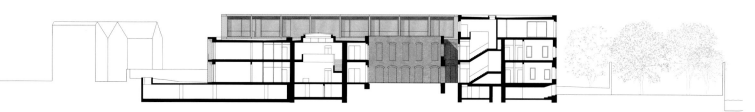

B-B' 剖面图 section B-B'

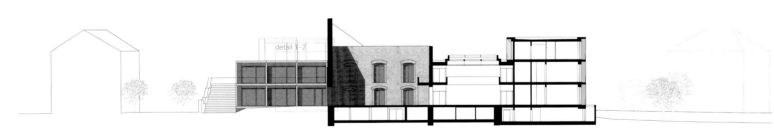

C-C' 剖面图 section C-C'

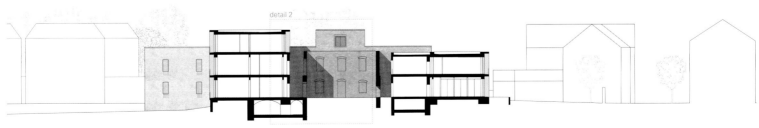

D-D' 剖面图 section D-D'

洛桑州立美术馆
Cantonal Museum of Fine Arts

Barozzi Veiga

庞大的美术馆将铁路和新的公共空间分隔开来
A monolithic museum separates rail lines and new public space

洛桑州立美术馆位于瑞士沃州的洛桑市,其长方形的建筑体量占地面积巨大,由砖块砌成。美术馆由巴塞罗那的Barozzi Veiga建筑事务所设计,是Plateforme 10(洛桑市的新艺术区)的主要建筑,由同一家事务所进行总体规划。美术馆建筑面积6895m²,轮廓采用矩形结构,长147.6m,宽22m,沿着该地区新建的长长的公共广场延伸,广场内还新建有另外两座美术馆。美术馆另外一侧纵向立面保护了这个区域不受火车噪声的干扰,紧邻的铁路线服务于附近的中央火车站,在美术馆的东部。

美术馆以其实用、严谨的几何形状与硬朗尖锐的线条与场地原有的工业特征相呼应。原有基础设施的要素得以留存下来,特别是原有机车库中的一部分,这个机车库从南立面向铁路线延伸。新建的全高度中央门厅成为设计的主要组成元素,原有的拱形窗户让阳光直射进来,通过拱形窗户下方宽阔的楼梯,有序地连接着三层楼的空间。

一层是由公共广场延伸出来的,主要的公共项目便位于这一层,包括咖啡馆、书店、礼堂和冥想空间,以及一个项目空间。这一层立面的门窗洞口比较多,从而让这些内部功能与外部公共空间连接起来。

展区位于更高楼层,在门厅的两侧。西侧存放的是永久馆藏,高4.5m,占地面积1750m²;而东侧是临时展区,高5.5m,占地面积1470m²。独立的垂直交通流线可容纳未来的综合性展览,更小一些的藏品也可以容纳进去。美术馆有五部楼梯。

顶层通过朝北的模块化天窗结构进行自然采光,并通过内部百叶窗系统来过滤和调节光线,进而为艺术品创造最佳的展示条件。

砖立面为混凝土框架增加了结构支撑,这一外观让人回忆起场地的工业历史,并为这座庞大的单体建筑提供了纹理以及生动活泼的图案。整体建筑相对封闭。为了保护藏品,美术馆有一个透明、坚实的南立面,除了保存的遗留结构外,只留有两扇窗户。东端实体立面的特征是嵌入了之前建筑的轮廓。北立面更加开放、通透和生动,它是由间隔规则的平行垂直石板组成的,这些石板有节奏的排列打破了巨石的厚重感,并露出了开口的位置。在夜晚,这些垂直石板反射美术馆内部的灯光,扩散开去。北立面的入口呈矩形,连接了公共空间和中央大厅。

建筑师认为该建筑可被定义为一面可居住的墙,这面墙将工业世界和新的公共空间分隔开,分毫不差。

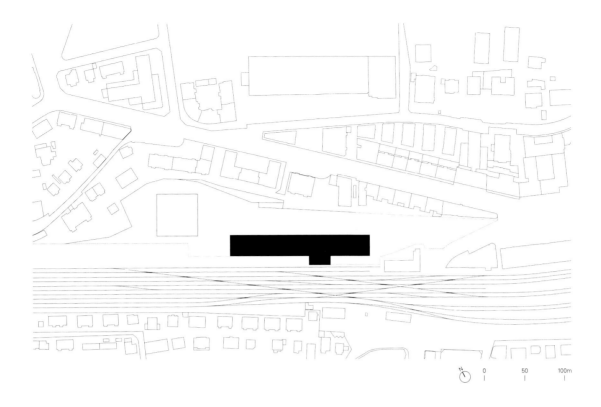

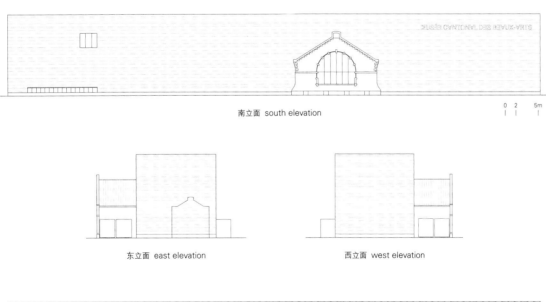

南立面 south elevation

东立面 east elevation　　　西立面 west elevation

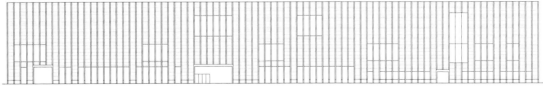

北立面 north elevation

项目名称：Musée cantonal des Beaux-Arts Lausanne / 地点：Place de la Gare 16, 1003 Lausanne, Switzerland / 建筑师：Barozzi Veiga
项目团队：execution phase_Pieter Janssens, Claire Afarian, Alicia Borchardt, Paola Calcavecchia, Marta Grzadziel, Isabel Labrador, Miguel Pereira Vinagre, Cristina Porta, Laura Rodriguez, Arnau Sastre, Maria Ubach, Cecilia Vielba, Nelly Vitiello, Alessandro Lussignoli / 竞赛阶段团队：competition phase_Roi Carrera, Shin Hye Kwang, Eleonora Maccari, Verena Recla, Agnieszka Samsel, Agnieszka Suchocka / 建筑承包商：Specific constructor for different construction elements / 项目经理：Pragma Partenaires SA / 结构顾问：Ingeni SA / 客户：Canton de Vaud, Direction générale des immeubles et du patrimoine, Architecture et Ingénierie cantonale (DGIP) / 建筑面积：6,895m² / 建成面积：12,449m²
造价：CHF 84,500,000 / 设计时间：2011 / 施工时间：2016—2019 / 竣工时间：2019 / 摄影师：©Simon Menges (courtesy of the architect)

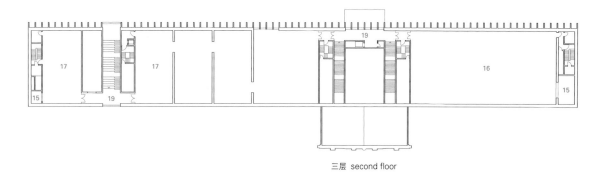

三层 second floor

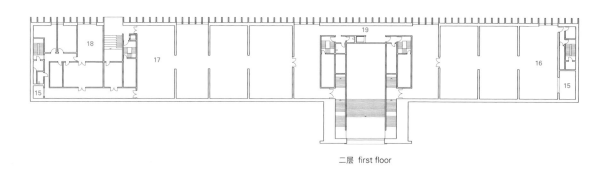

二层 first floor

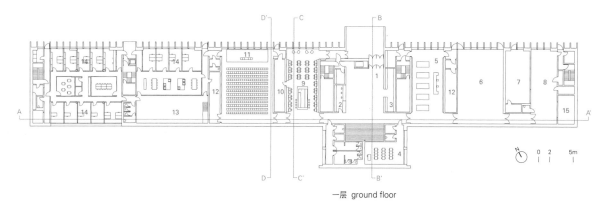

一层 ground floor

1. 入口大厅 2. 售票厅 3. 衣帽存放处 4. 教育工作室 5. 书店 6. 临时展厅 7. 美术馆工作坊 8. 装货区 9. 餐馆
10. 厨房 11. 礼堂 12. 储藏室 13. 图书馆 14. 办公室 15. 货运电梯 16. 临时展厅 17. 藏品 18. 保护实验室 19. 门厅
1. entrance hall 2. ticket office 3. cloakroom 4. education studio 5. bookshop 6. temporary exhibition 7. museum workshop 8. loading dock 9. restaurant
10. kitchen 11. auditorium 12. storage 13. library 14. offices 15. freight elevator 16. temporary exhibition 17. collection 18. conservation lab 19. foyer

The Cantonal Museum of Fine Arts occupies a long monolithic, rectilinear brick-clad volume in Lausanne, a city in the Swiss canton of Vaud. Designed by Barcelona-based Barozzi Veiga, it is the anchor building of Plateforme 10, the city's new art district, which was masterplanned by the same architect. The 6,895m² museum has a rectangular form, 147.6m long and 22m deep, stretching alongside the district's long new public plaza, where two other new museums gravitate. The other longitudinal side protects the district from train noise from the rail lines which serve the nearby central station to the east.

The museum references the memory of the formerly industrial site with pragmatic forms, rigorous geometry and hard, sharp lines. Elements of previous infrastructure are incorporated, notably a fragment of the former locomotive shed, which extends from the south facade towards the rail lines. Its original arched window brings natural light into the new full-height central foyer, which becomes the design's main compositional element, connecting the building's sequence of spaces over three floors via a wide staircase below the preserved window.

The ground floor is conceived as the extension of the public square and shelters the main public programs including a café, bookshop, auditorium and meditation space, as well as a project space. The facade on this level is very porous in order to allow for these internal functions to be continuous with the exterior public space.

Exhibition spaces are situated on the higher levels, on both sides of the foyer. The permanent collection occupies 1,750m² of on the west side, and 1,470m² of temporary galleries are to the east. Ceiling heights are 4.5m and 5.5m. Independent vertical circulations allow future comprehensive exhibitions as well as smaller capsule collections. The museum has five staircases.

The upper floor is naturally lit through north oriented modular skylight structures, which filter and adjust light coming through with an internal system of blinds so that they produce the optimal conditions for the art pieces.

The brick facades add structural support to a concrete frame. They evoke the industrial history of the site and offer texture and a vibrant pattern to the monolith. The overall building is relatively hermetic. In order to protect the collections, the museum has a sheer, solid southern facade punctuated by only two windows in addition to the retained heritage structure. The eastern end's solid facade is marked by a line tracing the section of the previous building. A more open, permeable and animated northern facade is created by regularly spaced parallel vertical fins, whose rhythm breaks the massiveness of the monolith and reveals the openings. At night, the fins diffuse the interior light coming from the museum. A rectangular entrance structure connecting the public space and the central foyer emerges from this facade.

The architect states that the building can be defined as an inhabited wall that separates, with precision, the industrial world from the new public space.

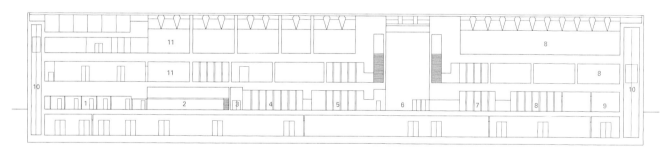

A-A' 剖面图 section A-A'

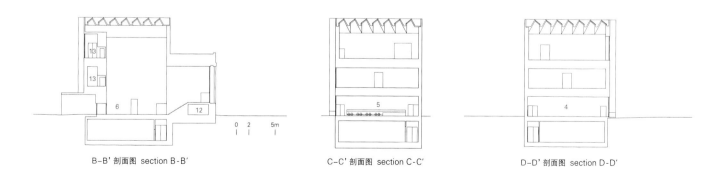

B-B' 剖面图 section B-B'   C-C' 剖面图 section C-C'   D-D' 剖面图 section D-D'

1. 办公室 2. 图书馆 3. 储藏室 4. 礼堂 5. 餐馆 6. 入口大厅 7. 书店 8. 临时展区 9. 装货区 10. 货运电梯 11. 藏品区 12. 教育工作室 13. 门厅
1. offices 2. library 3. storage 4. auditorium 5. restaurant 6. entrance hall 7. bookshop 8. temporary exhibition 9. loading dock 10. freight elevator 11. collection 12. education studio 13. foyer

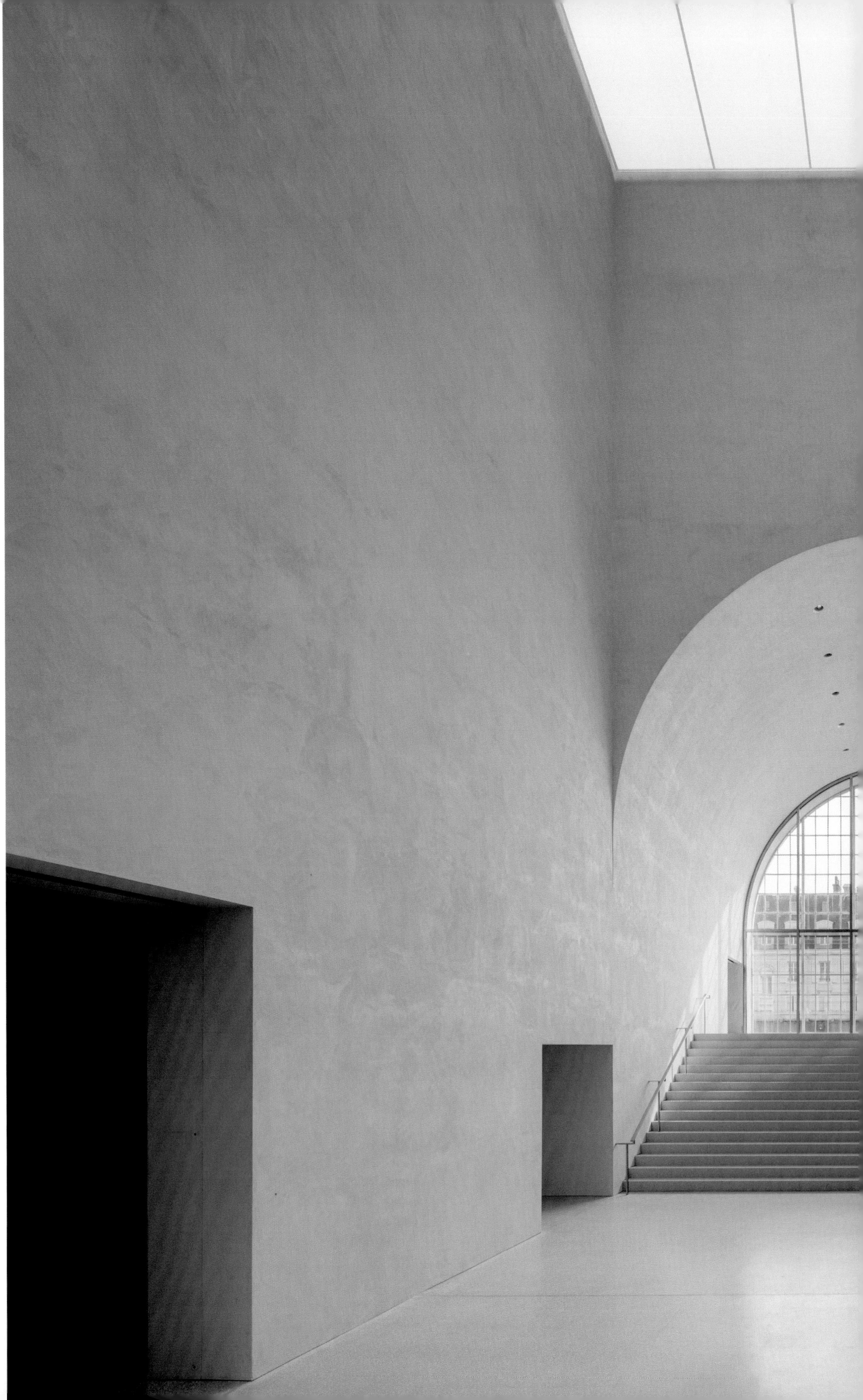

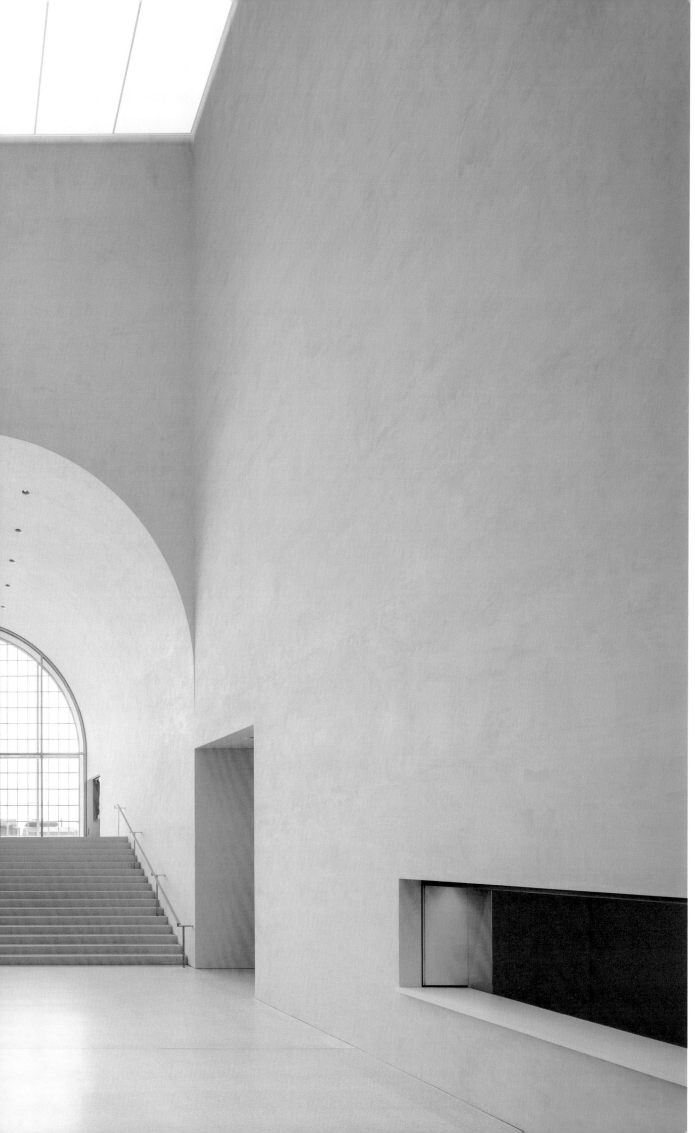

科隆学校附属建筑和住房
School Annex and Housing, Cologne - Lindenthal

LRO Lederer Ragnarsdóttir Oei Architekten

功能逻辑衍生出形式复杂的学校扩建
Functional logic generates a complex form in a school expansion

德国科隆一所学校综合楼的扩建由LRO (Lederer Ragnarsdóttir Oei Architekten) 建筑工作室设计，工作室位于斯图加特。扩建项目创造了一个由砖块砌筑而成的建筑体量，融入到其宁静的环境之中。该项目扩建了Liebfrauenschule中学 (斯图亚特圣母学校)、Domsingschule小学 (斯图亚特教堂唱诗班学校) 以及Kölner Dommusik学校 (科隆教堂音乐学校) 三所学校的校区，以及其教室、排练室和中央食堂。上层有十个房屋住房单元，用以补充学校的功能。

针对音乐培训，特别是针对斯图亚特教堂唱诗班学校来说，新的排练室可以提供个人教学和分组教学空间。斯图亚特教堂唱诗班学校和斯图亚特圣母学校都将教学活动向全日制学校靠拢，因此需要一个食堂。食堂的设计可以同时供300人用餐，用餐次数一天高达800人次。

此外，还有一种想法是：将食堂的用餐处作为活动场所。

教育园区的占地面积约为3300m²，位于基督复活教堂 (1970年) 的正南面。该教堂是一座标志性的野兽派建筑，由戈特弗里德·波姆设计。二者围绕在一个下沉的公园庭院的东侧。克拉伦巴赫大街向北延伸，成了前往教堂的道路。往东，沿着布鲁克纳大街有一座种着栗树的公园。建筑的西面毗邻面向教堂庭院的斯图亚特教堂唱诗班学校现有的体育馆。

新建的建筑综合体由地上四层和一个地下层组成。整体建筑由三个大小不同的体量组成。主体建筑通过中央大厅进入一层。该建筑的一侧是厨房，其中包括卫浴设施。另一侧是四个大型排练厅，可以从中央走廊进入。

上层有10个住房单元，每个单元都有80m²到140m²的生活空间。二层有30个小练习室供学生使用。地下层有一个地下停车场、技术设备区和储藏室，以及学校的卫生间设施。

与主体建筑南侧相连的是单层的食堂建筑。一个轻型木结构横跨整个食堂区域，支撑着屋顶轻巧的组件。

第三个建筑体量有两层直接通向基督复活教堂。人们可以穿过一个朝向斯图亚特教堂唱诗班学校的庭院 (歌唱花园) 进入教室和食堂，也可以通过一条通道进入教堂的前院，而从北面的克拉伦巴赫大街便可前往住房单元。

砖立面是德国城市结构的一个特色部分，但本项目的独特之处在于：不同寻常的各种建筑外形混合在一起，十分有趣。它与相邻的基督复活教堂相呼应，后者的特点是：下层的砖块和上层的混凝土表现出复杂的建筑形式。学校附属建筑是对波姆所选用的砖块材质以及大胆的雕塑风格的回应。LRO建筑工作室设计的附属建筑造型各异，例如：上层住宅楼层采用了不同的悬臂结构；长长的弧形窗户穿过一面普通的墙壁，从餐厅可以直接看到公园绿地；每一面最长的墙壁都有一处令人惊奇的弧形凹口；还有礼堂半圆形座位的平面布局。该建筑虽然有些复杂，但却是根据功能逻辑建造出来的。其低调内敛的设计使建筑具有永恒的存在感，完美融入到科隆安静的林登塔尔街区当中。

The expansion of a school complex in Cologne, Germany by Stuttgart-based architects LRO has created a composition of connected brick volumes which assimilates into its tranquil setting. The project expanded the premises of the Liebfrauenschule (School of Our Lady) secondary school, the Domsingschule (Cathedral Choir School) primary school, and the Kölner Dommusik (Cologne Cathedral Music) with additional classrooms and rehearsal rooms as well as a central canteen. The school functions are complemented by ten housing units on the upper floors.

For the musical training, especially at the Domsingschule, new rehearsal rooms have been provided for individual and group instruction. Both the Domsingschule and the Liebfrauenschule have oriented their teaching activities to an all-day school program, resulting in the need for a canteen. It is designed with the capacity to serve up to 300 meals simultaneously and up to 800 meals in total each day. Furthermore, the intention is to be able to use the canteen's dining hall as an event venue. The building plot measures approx. 3,300m². It is located directly south of the Church of the Resurrection of Christ (1970), an iconic brutalist building designed by Gottfried Böhm, and together they enclose the eastern side of a sunken park courtyard. Clarenbachstrasse runs along the north and provides access to the property. To the east, along Brucknerstrasse, is a

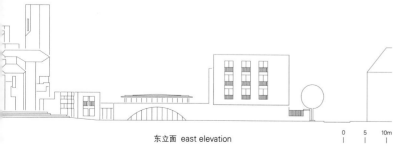

东立面 east elevation

北立面 north elevation

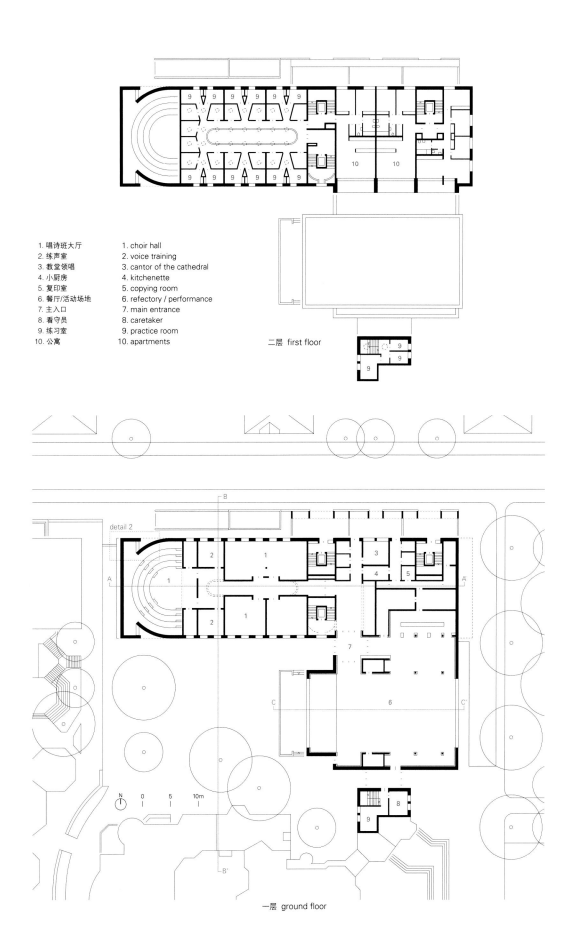

1. 唱诗班大厅　　1. choir hall
2. 练声室　　　　2. voice training
3. 教堂领唱　　　3. cantor of the cathedral
4. 小厨房　　　　4. kitchenette
5. 复印室　　　　5. copying room
6. 餐厅/活动场地　6. refectory / performance
7. 主入口　　　　7. main entrance
8. 看守员　　　　8. caretaker
9. 练习室　　　　9. practice room
10. 公寓　　　　 10. apartments

二层 first floor

一层 ground floor

项目名称：School Annex and Housing, Cologne-Lindenthal / 地点：Clarenbachstraße 1, Cologne-Lindenthal, Germany / 建筑师：LRO Lederer Ragnarsdóttir Oei Architekten / 项目团队：Frank Bohnet, Philipp Gantenbrink / 项目管理：Drees＆Sommer Köln GmbH / 结构工程：Leonhardt, Andrä und Partner Beratende Ingenieure VBI AG / 工程建设测试统计数据：Pirlet＆Partner Baukonstruktionen, Ingenieurgesellschaft mbH / 建设工程：Ingenieurbüro Heiming / 结构物理：ISRW Klapdor GmbH / 消防设计：Kempen Krause Ingenieure GmbH / 客户：Erzbistum Köln, Generalvikariat Abteilung Bau / 总建筑面积：5,300m² / 有效面积：3,800m² / 施工时间：2017—2020
摄影师：©Roland Halbe

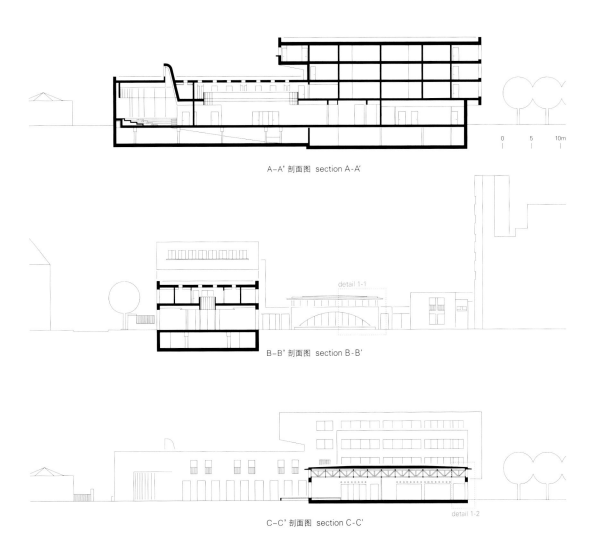

A-A' 剖面图 section A-A'

B-B' 剖面图 section B-B'

C-C' 剖面图 section C-C'

public park with chestnut trees. To the west the property borders on an existing gymnasium for the Domsingschule facing the courtyard.

The building complex is a new construction consisting of three upper floors, the ground floor and a basement. The building form is articulated into three different volumes. The principal volume is entered on the ground floor via a central lobby. Adjoining one side of this volume is the kitchen, including sanitary facilities. On the other side are four large rehearsal halls, which are accessed from a central corridor.

On the upper stories are ten housing units, each with approximately 80 to 140m² of living space. The first floor is supplemented by 30 practice booths for school use. In the basement are an underground car park, technical and storage areas and the toilet facilities for the school.

Connected to the southern side of the main volume is the single-story block of the canteen dining room. This area is spanned by a lightweight wood structure supporting the lightest possible roof assembly.

The third volume directly adjoins the Church of the Resurrection over two stories. The classrooms and canteen are accessible across a courtyard (the "Sing Garden") that opens to the Domsingschule and through a passageway to the forecourt of the church, while access to the housing units is from the north via Clarenbachstrasse.

Facades of brick are a characteristic part of German city fabric, but this project is distinctive in its unusual and playful variety of incorporated shapes. It offers an echo of the adjacent Church of the Resurrection of Christ, which is characterized by its complex form expressed in brick at lower levels and concrete above. The school annex responds to Böhm's brick materiality and sculptural adventure. The variety of shape in LRO's annex includes different cantilevers in the upper residential floors, a plain wall penetrated by a long arc of window which reveals parkland to the canteen, and an unexpected curved indent in each of the longest walls, following the semi-circular seating plan of an auditorium inside. The building is complex, but is shaped by functional logic. Its restrained materiality gives it a timeless presence suitable to Cologne's quiet neighborhood of Lindenthal.

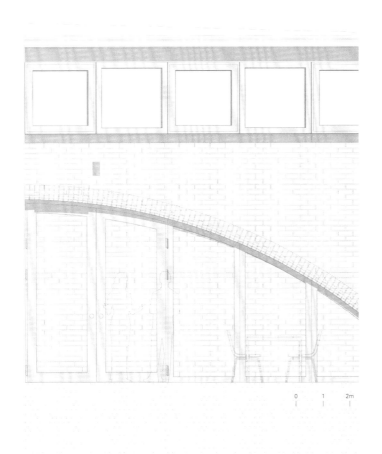

详图1-1 detail 1-1

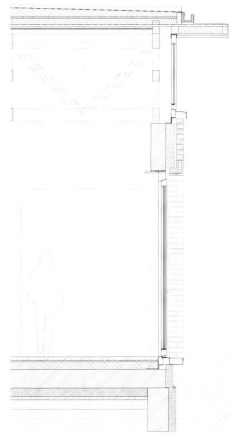

详图1-2 detail 1-2

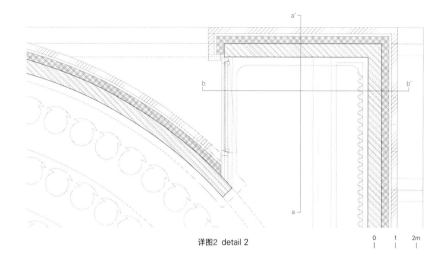

详图2 detail 2

0 1 2m

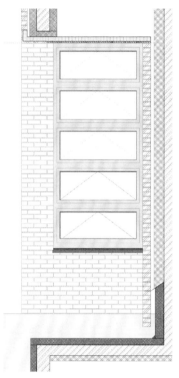

a-a' 剖面详图 detail a-a'

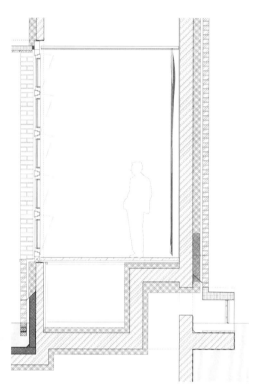

b-b' 剖面详图 detail b-b'

Doreen Heng LIU

P94 NODE Architecture & Urbanism

Doreen Heng LIU, born in Guangzhou, received a MArch from UC Berkeley and a Doctorate from Harvard Graduate School of Design. Since 2008, she has been teaching at the School of Architecture, Chinese University of Hong Kong as adjunct associate professor. Her research focuses on contemporary urbanism and architecture in the Pearl River Delta (Greater Bay Area) & China, and their specific impacts in the making of architecture today. In 2004, she established her own design practice NODE in Nansha and Hong Kong, and relocated in Shenzhen since 2009. With her design focuses on urban regeneration, infrastructure & public space, her studio tries to re-investigate & re-examine the given conditions based on specific sites and issues. In 2014, Doreen was nominated as Curator for Hong Kong Pavilion for the 14th International Architecture Exhibition – la Biennale di Venezia. In 2015, Doreen was nominated as chief curator for Shenzhen Hong Kong Urbanism/Architecture Bi-city Biennial, together with Aaron Betsky, Hubert Klumpner & Alfredo Brillembourg.

P110 Henning Larsen

Since its founding 60 years ago, Henning Larsen has worked from a simple foundational ethos: people come first. A deep curiosity for the world has guided the Danish studio to adapt their approach to Nordic modernism across countless contexts and communities; from the electric cityscapes of Hong Kong and New York to the rugged slopes of the Faroe Islands. The office, now 300-strong, brings its unique perspective to projects and contexts across the world, working from offices in Copenhagen, New York, Hong Kong, Munich, Oslo, and the Faroe Islands.

Instead of looking at architecture as a stand-alone object, the studio treats it as a medium formed for and by the social and environmental systems in and around it. This has deep roots in the Scandinavian design tradition, in which aesthetic, social, and intellectual design features are equally considered to develop vibrant and contextual work that reaches beyond itself.

©Peter Norby

P4 Adrian Friend

Adrian Friend is an RIBA Educator and has taught and lectured throughout the UK. Credited with establishing a projects office for University of Nottingham and leading the inaugural Jouberton Nursery 'live-project', designed and built by students in South Africa, Adrian has also helped implement a new architecture undergraduate programme for Norwich University of the Arts, returning the study of architecture to the city after a 50 year absence. A champion of design quality Adrian writes for the leading design and architecture publications including *Blueprint*, *Wallpaper* and *Building Design*.

P204 LRO Lederer Ragnarsdóttir Oei Architekten

The office LRO was founded in 1979 by Arno Lederer[center]. Followed by an office partnership with Jórunn Ragnarsdóttir[right] since 1985 and at last followed by the office partnership with Marc Oei[left] since 1992. LRO currently has about 50 employees, and is based in Stuttgart, Germany. For the entire duration of the work process LRO works in small teams to generate a high level of intensity and identification with the projects. In the recent years they won the competitions including the Annex to the Württemberg State Library in Stuttgart, the School Annex and Housing in Cologne-Lindenthal, the Historical Museum in Frankfurt a. M., the Municipal Museum in Stuttgart, and the Diocesan Curia and archive in Rottenburg.
Their projects are mainly realized in the fields: office buildings, administrative buildings, leisure and sports, health and education, cultural facilities, church, housing, industry as well as urban planning.

©Ingrid von Kruse

P170 David Chipperfield Architects

Founded in London in 1985, has won numerous international competitions and built over 100 projects worldwide. Its diverse body of work includes cultural, residential, commercial and educational buildings, as well as civic projects and urban masterplans. Offices in London, Berlin, Milan and Shanghai contribute to the wide range of projects and typologies.
The practice's work is characterised by meticulous attention to the concept and details of every project, and a relentless focus on refining design ideas to arrive at a solution which is architecturally, socially, environmentally and intellectually coherent. Has won more than 50 international awards and citations for design excellence, including the RIBA Stirling Prize for The Museum of Modern Literature in Marbach, Germany, and both the Mies van der Rohe Award and the Deutscher Architekturpreis for the Neues Museum in Berlin.

©Zooey Braun

P66 Hyungmin Pai

Architectural historian, critic, and curator, is currently Professor at the University of Seoul. Twice a Fulbright Scholar, he studied architecture and urban design at Seoul National University and received his Ph.D from MIT. Is author of *The Portfolio and the Diagram* (2002), *Sensuous Plan: The Architecture of Seung H-Sang* (2007), *Key Concepts of Korean Architecture* (2013), *Architecture of Amorepacific* (2018), and *Doubt is Power* (2019). For the Venice Biennale, he was twice curator for the Korean Pavilion, which in 2014, was awarded the Golden Lion. He was Director of the inaugural Seoul Biennale of Architecture and Urbanism 2019 and was appointed as curator for major international exhibitions.

P124 Oppenheim Architecture

Founded by Chad Oppenheim in 1999, Oppenheim Architecture is an architecture, planning, and interior design firm specializing in hospitality, commercial mixed-use, retail and residential buildings worldwide. Has received over 80 awards and distinctions, including the AIA's highest distinction, the Silver Medal, and featured in numerous publications around the world. The studio's work is built on the belief that design follows life and that form follows feeling; crafting projects that are as beautiful as it is functional. The firm designs with sensitivity toward man and nature – harmonizing with the surroundings of each context.
Chad Oppenheim, a Miami-based architect, graduated from Cornell University. Has taught at various architecture schools including Harvard University's Graduate School of Design and most recently at Cornell University's College of Architecture, Art, and Planning.

P78 BIG

Was founded in 2005 by Bjarke Ingels, BIG is a Copenhagen, New York and London based group of architects, designers, urbanists, landscape professionals, interior and product designers, researchers, and inventors. Currently involves in a large number of projects throughout Europe, North America, Asia, and the Middle East. Believes that in order to deal with today's challenges, architecture can profitably move into a field that has been largely unexplored. A pragmatic utopian architecture that steers clear of the petrifying pragmatism of boring boxes and the naïve utopian ideas of digital formalism. Like a form of programmatic alchemist, it creates architecture by mixing conventional ingredients such as living, leisure, working, parking, and shopping. By hitting the fertile overlap between pragmatic and utopia, once again finds the freedom to change the surface of our planet, to better fit contemporary life forms.

P188 Barozzi Veiga

Was founded in Barcelona by Fabrizio Barozzi[left] and Alberto Veiga[right] in 2004. Barozzi Veiga's work, which mainly includes cultural and educational buildings, is characterized by the intention to arrive at solutions that are rooted in place, architectures that can be perceived over time and that have an emotional content. Concepts and ideas which are able to create particular atmospheres, that are architecturally clear and expressive, and able to have a meaning by itself.
Barozzi Veiga won numerous national and international prizes: Ajac Young Catalan Architect Award (2007), Barbara Cappochin International Architecture Award (2011), Gold Medal for Italian Architecture for the Best Debut Work with Ribera del Duero Headquarter (2012), Mies van der Rohe Award for European Architecture (2015) with the project for the Szczecin Philharmonic, Chicago Atheneum International Award (2019), AD Award Architects of the Year (2019). Has been invited to contribute to several international exhibitions, including the Chicago Architecture Biennal (2017) and the Biennale di Venezia (2014 and 2016).

P142 Richard Ingersoll

Born in California, 1949, earned a doctorate in architectural history at UC Berkeley, and was a tenured associate professor at Rice University (Houston) form 1986-97. Has lived off and on in Tuscany since 1970 and currently teaches at Syracuse University in Florence (Italy), and the Politecnico in Milan. He was the executive editor of *Design Book Review* from 1983 to 1997. His recent publications iunclude: *World Architecture, A Cross-Cultural History* (2013); *Sprawltown, Looking for the City on its Edge* (2006); *World Architecture* (1900-2000); *A Critical Mosaic, Volume 1: North America, USA and Canada* (2000). Frequently writes criticism for *Arquitectura Viva, Architect, Lotus* and *C3*.

P8 Seung H-Sang

Born in 1952, graduated from Seoul National University and studied at Vienna University of Technology. Worked for Kim Swoo-geun from 1974 to 1989 and established IROJE architects & planners in 1989. Taught at North London University (1998-1997), Seoul National University (2003-2005), Korea National University of Art (2007-2009), TU in Vienna (2017-2018) and CAFA in Beijing (2019-2020). Is now chair professor of Donga University since 2018.
Is author of *Beauty of Poverty* (1996), *Architecture, Signs of Thoughts* (2004), *Landscript* (2009), *All Oldies are Beautiful* (2012), *Invisible Architecture, Inconstant Cities* (2016), *Meditation* (2019) and *Natured* (2020).
The America Institute of Architects invested him with Honorary Fellow of AIA in 2002, and Korea National Museum of Contemporary of Art selected him as 'Artist of the Year 2002'. In 2019 the Austrian Government awarded him 'Cross of Honour for Science and Art, First Class'. Korean government who honored him with 'Korea Award for Art and Culture' in 2007, again invested him with 'Order of Cultural Merit, Eun-gwan' in 2020.

P70 Nelson Mota

Graduated at the University of Coimbra, and teaches architectural design and theory at Delft University of Technology. In his doctoral dissertation "An Archaeology of the Ordinary" (Delft University of Technology, 2014) he examined the relationship between housing design and the reproduction of vernacular social and spatial practices. Is production editor and member of the editorial board of *Footprint* and a founding partner of Comoco architects.

Francesca Torzo

P148 Francesca Torzo Architetto

Francesca Torzo studied in TU Delft and ETSAB Barcelona. Received a diploma at AAM in Mendrisio and graduated with honours by IUAV in Venezia in 2001 with arch. P. Zumthor, prof. U. Tubini, arch. M. Kreisler, eng. J. Conzett. During 2001-2002 she worked as project architect of Peter Zumthor Architekturburo. In 2008 she started her own office in Genova. Lectured at a number of schools and cultural institutions, such as Ecole Nationale Supérieure des Arts de Paris Cergy, AUT Innsbruck, Technische Universität München, USI Accademia di Architettura di Mendrisio, Technische Universität Wien, Museum Alvar Aalto Helsinki, Triennale di Milano, and the Architecture Foundation London. In 2020 she is awarded the WA Moira Gemmill prize. The studio is a small international team of young architects. The design process starts from an understanding of the material constrains and of the cultural context. The goal is that of formulating the primary spatial relations, which have to be preserved as a narrative through the complete process. They focus on design by pursuing a vision that embraces the world in which we live and its memories, based on sympathy for the contradictions in human life.

墙体设计
ISBN: 978-7-5611-6353-5
定价: 150.00 元

新公共空间与私人住宅
ISBN: 978-7-5611-6354-2
定价: 150.00 元

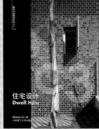
住宅设计
ISBN: 978-7-5611-6352-8
定价: 150.00 元

老年住宅
ISBN: 978-7-5611-6569-0
定价: 150.00 元

小型建筑
ISBN: 978-7-5611-6579-9
定价: 150.00 元

文博建筑
ISBN: 978-7-5611-6568-3
定价: 150.00 元

流动的世界：日本住宅空间设计
ISBN: 978-7-5611-6621-5
定价: 200.00 元

创意运动设施
ISBN: 978-7-5611-6636-9
定价: 180.00 元

墙体与外立面
ISBN: 978-7-5611-6641-3
定价: 180.00 元

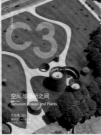
空间与场所之间
ISBN: 978-7-5611-6650-5
定价: 180.00 元

文化与公共建筑
ISBN: 978-7-5611-6746-5
定价: 160.00 元

城市扩建的四种手法
ISBN: 978-7-5611-6776-2
定价: 180.00 元

复杂性与装饰风格的回归
ISBN: 978-7-5611-6828-8
定价: 180.00 元

企业形象的建筑表达
ISBN: 978-7-5611-6829-5
定价: 180.00 元

图书馆的变迁
ISBN: 978-7-5611-6905-6
定价: 180.00 元

亲地建筑
ISBN: 978-7-5611-6924-7
定价: 180.00 元

旧厂房的空间蜕变
ISBN: 978-7-5611-7093-9
定价: 180.00 元

混凝土语言
ISBN: 978-7-5611-7136-3
定价: 228.00 元

建筑入景
ISBN: 978-7-5611-7306-0
定价: 228.00 元

新医疗建筑
ISBN: 978-7-5611-7328-2
定价: 228.00 元

内在丰富性建筑
ISBN: 978-7-5611-7444-9
定价: 228.00 元

建筑谱系传承
ISBN: 978-7-5611-7461-6
定价: 228.00 元

伴绿而生的建筑
ISBN: 978-7-5611-7548-4
定价: 228.00 元

大地的皱折
ISBN: 978-7-5611-7649-8
定价: 228.00 元

在城市中转换
ISBN: 978-7-5611-7737-2
定价: 228.00 元

锚固与飞翔——挑出的住居
ISBN: 978-7-5611-7759-4
定价: 228.00 元

创造性加建：我的学校，我的城市
ISBN: 978-7-5611-7848-5
定价: 228.00 元

文化设施：设计三法
ISBN: 978-7-5611-7893-5
定价: 228.00 元

终结的建筑
ISBN: 978-7-5611-8032-7
定价: 228.00 元

博物馆的变迁
ISBN: 978-7-5611-8226-0
定价: 228.00 元

微工作·微空间
ISBN: 978-7-5611-8255-0
定价: 228.00 元

居住的流变
ISBN: 978-7-5611-8328-1
定价: 228.00 元

本土现代化
ISBN: 978-7-5611-8380-9
定价: 228.00 元

气候与环境
ISBN: 978-7-5611-8501-8
定价: 228.00 元

能源与绿色
ISBN: 978-7-5611-8911-5
定价: 228.00 元

体验与感受：艺术画廊与剧院
ISBN: 978-7-5611-8914-6
定价: 228.00 元

记忆的住居
ISBN: 978-7-5611-9027-2
定价: 228.00 元

场地、美学和纪念性建筑
ISBN: 978-7-5611-9095-1
定价: 228.00 元

殡仪类建筑：在返璞和升华之间
ISBN: 978-7-5611-9110-1
定价: 228.00 元

苏醒的儿童空间
ISBN: 978-7-5611-9182-8
定价: 228.00 元

都市与社区
ISBN: 978-7-5611-9365-5
定价: 228.00 元

木建筑再生
ISBN: 978-7-5611-9366-2
定价: 228.00 元

© 2021大连理工大学出版社

版权所有·侵权必究

图书在版编目(CIP)数据

建筑重现 / 丹麦BIG建筑事务所等编；司炳月，李一同，孙彤彤译. — 大连 ：大连理工大学出版社，2021.5
　ISBN 978-7-5685-2980-8

　Ⅰ．①建… Ⅱ．①丹… ②司… ③李… ④孙… Ⅲ.①建筑设计 Ⅳ．①TU2

中国版本图书馆CIP数据核字(2021)第068058号

出版发行：大连理工大学出版社
　　　　（地址：大连市软件园路80号　邮编：116023）
印　　刷：上海锦良印刷厂有限公司
幅面尺寸：225mm×300mm
印　　张：14
出版时间：2021年5月第1版
印刷时间：2021年5月第1次印刷
统　　筹：房　磊
责任编辑：张昕焱
封面设计：王志峰
责任校对：杨　丹
书　　号：978-7-5685-2980-8
定　　价：298.00元

发　行：0411-84708842
传　真：0411-84701466
E-mail：12282980@qq.com
URL：http://dutp.dlut.edu.cn

本书如有印装质量问题，请与我社发行部联系更换。